Praise for *Programming Your Home*

Mike has a broad technology experience base that puts all the pieces of some remarkable projects together. It's amazing that he makes it all so easy and affordable. Don't miss all that can be learned from this gem.
➤ Michael Bengtson, Consultant

The Web-Enabled Light Switch project gave my family convenience and security options and enhanced my knowledge of RS-232 communications. It is nice to be able to switch on lights from my favorite chair. And the Tweeting Bird Feeder project has opened my eyes to the uses of radio communications around the home for things besides Wi-Fi, and it will help in my work to contribute to the preservation of bird species that are struggling for food and habitat.
➤ Bob Cochran, Information Technology Specialist

With this book, Mike Riley celebrates the Arduino microcontroller in a way that both beginning and advanced home automation hobbyists will enjoy.
➤ Sven Davies, Vice President of Applications

This is an outstanding reference that should be on the desk of every DIYer. In much the same way that software engineers mention "The Gang of Four Patterns Book," I predict this text will eventually be referred to as "The Riley Book of Home Automation."
➤ Jon Kurz, President, Dycet, LLC

Every technology is only as exciting as the things you do with it. Mike takes a few cheap electronics parts, an Arduino, and a bit of code and turns your home into a much more exciting and enjoyable place. His easy-to-follow instructions make every single one of these projects both fun and useful.
➤ **Maik Schmidt, Software Developer, Author of** *Arduino: A Quick-Start Guide*

I've had more fun learning new languages, systems, and gadgets with this book than any other book I've read!
➤ **James Schultz, Software Developer**

Home automation is great fun, and *Programming Your Home* by Mike Riley will get you started right away. By leveraging this book and the easily available free/inexpensive hardware and software, anyone can tackle some great projects.
➤ **Tony Williamitis, Senior Embedded Systems Engineer**

This is a fun and enthusiastic survey of electronic devices that can interact with the real world and that starts in your own home!
➤ **John Winans, Chief Software Architect**

Programming Your Home
Automate with Arduino, Android, and Your Computer

Mike Riley

The Pragmatic Bookshelf

Dallas, Texas • Raleigh, North Carolina

Many of the designations used by manufacturers and sellers to distinguish their products are claimed as trademarks. Where those designations appear in this book, and The Pragmatic Programmers, LLC was aware of a trademark claim, the designations have been printed in initial capital letters or in all capitals. The Pragmatic Starter Kit, The Pragmatic Programmer, Pragmatic Programming, Pragmatic Bookshelf, PragProg and the linking *g* device are trademarks of The Pragmatic Programmers, LLC.

Every precaution was taken in the preparation of this book. However, the publisher assumes no responsibility for errors or omissions, or for damages that may result from the use of information (including program listings) contained herein.

Our Pragmatic courses, workshops, and other products can help you and your team create better software and have more fun. For more information, as well as the latest Pragmatic titles, please visit us at *http://pragprog.com*.

The team that produced this book includes:

Jackie Carter (editor)
Potomac Indexing, LLC (indexer)
Molly McBeath (copyeditor)
David J Kelly (typesetter)
Janet Furlow (producer)
Juliet Benda (rights)
Ellie Callahan (support)

Copyright © 2012 The Pragmatic Programmers, LLC.
All rights reserved.

No part of this publication may be reproduced, stored in a retrieval system, or transmitted, in any form, or by any means, electronic, mechanical, photocopying, recording, or otherwise, without the prior consent of the publisher.

Printed in the United States of America.
ISBN-13: 978-1-93435-690-6
Printed on acid-free paper.
Book version: P1.0—February 2012

This book is dedicated to Bill, Eileen, and Josie.

Contents

Acknowledgments xi

Preface xiii

Part I — Preparations

1. Getting Started 3
   - 1.1 What Is Home Automation?   3
   - 1.2 Commercial Solutions   4
   - 1.3 DIY Solutions   5
   - 1.4 Justifying the Investment   5
   - 1.5 Setting Up Your Workbench   6
   - 1.6 Sketching Out Your Ideas   7
   - 1.7 Writing, Wiring, and Testing   8
   - 1.8 Documenting Your Work   9

2. Requirements 11
   - 2.1 Knowing the Hardware   12
   - 2.2 Knowing the Software   17
   - 2.3 Be Safe, Have Fun!   18

Part II — Projects

3. Water Level Notifier 23
   - 3.1 What You Need   23
   - 3.2 Building the Solution   25
   - 3.3 Hooking It Up   26
   - 3.4 Sketching Things Out   27
   - 3.5 Writing the Web Mailer   33
   - 3.6 Adding an Ethernet Shield   35

	3.7	All Together Now	39
	3.8	Next Steps	41
4.	**Electric Guard Dog**	**43**
	4.1	What You Need	44
	4.2	Building the Solution	45
	4.3	Dog Assembly	46
	4.4	Dog Training	50
	4.5	Testing It Out	53
	4.6	Unleashing the Dog	54
	4.7	Next Steps	55
5.	**Tweeting Bird Feeder**	**57**
	5.1	What You Need	57
	5.2	Building the Solution	60
	5.3	The Perch Sensor	61
	5.4	The Seed Sensor	65
	5.5	Going Wireless	68
	5.6	Tweeting with Python	73
	5.7	Putting It All Together	81
	5.8	Next Steps	82
6.	**Package Delivery Detector**	**85**
	6.1	What You Need	86
	6.2	Building the Solution	88
	6.3	Hardware Assembly	89
	6.4	Writing the Code	90
	6.5	The Package Delivery Sketch	90
	6.6	Testing the Delivery Sketch	92
	6.7	The Delivery Processor	93
	6.8	Creating the Delivery Database	93
	6.9	Installing the Package Dependencies	95
	6.10	Writing the Script	96
	6.11	Testing the Delivery Processor	100
	6.12	Setting It Up	102
	6.13	Next Steps	103
7.	**Web-Enabled Light Switch**	**105**
	7.1	What You Need	105
	7.2	Building the Solution	108
	7.3	Hooking It Up	109

	7.4	Writing the Code for the Web Client	112
	7.5	Testing Out the Web Client	114
	7.6	Writing the Code for the Android Client	115
	7.7	Testing Out the Android Client	119
	7.8	Next Steps	122
8.	**Curtain Automation**		**125**
	8.1	What You Need	125
	8.2	Building the Solution	128
	8.3	Using the Stepper Motor	129
	8.4	Programming the Stepper Motor	130
	8.5	Adding the Sensors	131
	8.6	Writing the Sketch	132
	8.7	Installing the Hardware	137
	8.8	Next Steps	140
9.	**Android Door Lock**		**141**
	9.1	What You Need	141
	9.2	Building the Solution	144
	9.3	Controlling the Android Door Lock	147
	9.4	Writing the Android Server	152
	9.5	Writing the Android Client	163
	9.6	Test and Install	167
	9.7	Next Steps	168
10.	**Giving Your Home a Voice**		**171**
	10.1	What You Need	171
	10.2	Speaker Setup	173
	10.3	Giving Lion a Voice	175
	10.4	Wireless Mic Calibration	177
	10.5	Programming a Talking Lion	179
	10.6	Conversing with Your Home	188
	10.7	Next Steps	189

Part III — Predictions

11.	**Future Designs**		**193**
	11.1	Living in the Near	193
	11.2	The Long View	196
	11.3	The Home of the Future	198

12. **More Project Ideas** **201**
    12.1 Clutter Detector   201
    12.2 Electricity Usage Monitor   202
    12.3 Electric Scarecrow   202
    12.4 Entertainment System Remote   202
    12.5 Home Sleep Timer   203
    12.6 Humidity Sensor-Driven Sprinkler System   203
    12.7 Networked Smoke Detectors   203
    12.8 Proximity Garage Door Opener   204
    12.9 Smart HVAC Controller   205
    12.10 Smart Mailbox   205
    12.11 Smart Lighting   205
    12.12 Solar and Wind Power Monitors   205

Part IV — Appendices

A1. **Installing Arduino Libraries** **209**
    A1.1 Apple OSX   209
    A1.2 Linux   210
    A1.3 Windows   210

A2. **Bibliography** **211**

   Index 213

Acknowledgments

I have been a lifelong tinkerer. My earliest recollection of dissecting my father's broken tape recorder instilled an appreciation for the technology that drove it. From there, erector sets, model railroads, and programmable calculators led to personal computers, mobile devices, and microcontrollers. Over the years, this passion for learning not only how stuff works but also how technical concepts can be remixed with surprising, often highly satisfying results has been liberating. That's why this book was such a joy for me to write.

Helping others to see what's possible by observing their surroundings and having the desire to take an active role in making their lives easier with technology while having fun is this book's primary goal. Yet without others helping me distill my ideas into what you are reading now, this book would not have been possible. It is to them that I wish to express my deepest gratitude for their support.

A boatload of thanks goes to the book's editor, Jackie Carter, who spent countless hours ensuring that my words were constructed with clarity and precision. Copy editor Molly McBeath did a fantastic job catching hidden (from my view anyway) typos and grammatical misconstructions. Big thanks to Susannah Pfalzer for her infectious enthusiasm and boundless boosts of encouragement and to Arduino expert and fellow Pragmatic author Maik Schmidt, whose own success helped pave the way for a book like this.

Many thanks also go to John Winans, tech wiz extraordinaire, who refactored the state machine code used in several of the projects, as well as to Sven Davies, Mike Bengtson, Jon Bearscove, Kevin Gisi, Michael Hunter, Jerry Kuch, Preston Patton, and Tony Williamitis for helping to make this book as technically accurate and complete as it is. Shout-outs also go to Jon Erikson and Jon Kurz for their enthusiastic encouragement. I also want to thank Bob Cochran and Jim Schultz for providing wonderfully helpful feedback during the book's beta period. Thanks also go to Philip Aaberg for filling my ears with music to code by. And to the makers of and contributors to the Arduino and Fritzing projects, you people have changed the world for the better.

I am most grateful to my wife, Marinette, and my family for allowing me to tunnel away for months in my mythical man cave to complete this book. And I can't gush enough over the wonderful pencil illustrations that my daughter drew for the book. I am so proud of you, Marielle!

Finally, I am sincerely thankful to Dave Thomas and Andy Hunt for their passion and vision. You're the best.

Mike Riley
`mailto:mike@mikeriley.com`
Naperville, IL, December 2011

Preface

Welcome to the exciting, empowering world of home automation! If you have ever wanted your home to do more than just protect you against the outside elements and want to interface it to the digital domain, this book will show you how. By demonstrating several easy-to-build projects, you will be able to take the skills you learned from this book and expand upon and apply them toward custom home automation projects of your own design.

The book's primary objective is to get you excited about the broader possibilities for home automation and instill the confidence you need to ultimately build upon these and your own ideas. The projects also make great parent-child learning activities, as the finished products instill a great sense of accomplishment. And who knows? Your nifty home automation creations may even change the world and become a huge new business opportunity for other homeowners actively seeking an automation solution that saves them time and money.

Who Should Read This Book

Programming Your Home is best suited to DIYers, programmers, and tinkerers who enjoy spending their leisure time building high-tech solutions to further automate their lives and impress their friends and family with their creations. Essentially, it is for those who generally enjoy creating custom technology and electronics solutions for their own personal living space.

A basic understanding of Arduino and programming languages like Ruby and Python are recommended but not required. You will learn how to combine these technologies in unique configurations to resolve homemaker annoyances and improve home management efficiencies.

In addition to the inclusion of Python scripts and Ruby on Rails-based web services, several of the projects call upon Google's Android platform to help enhance the data event collection, visualization, and instantiation of activities.

A basic familiarity with the Android SDK will be beneficial so that the projects that make use of the Android OS can offer a more mobile reach.

If you're the type of person who prefers to build versus buy your home accessories, this book will further motivate you to use what you learned in the book as a starting point to expand upon and optimize them in various ways for their environment. Even though some of the topics deal with multiple software- and hardware-based solutions, they are easy to follow and inexpensive to build. Most of all, they show how a few simple ideas can transform a static analog environment into a smart digital one while having fun.

What's in This Book

After a basic introduction to home automation and the tools of the trade, this book will teach you how to construct and program eight unique projects that improve home utility and leisure-time efficiencies. Each project incorporates a variety of inexpensive sensors, actuators, and microcontrollers that have their own unique functions. You will assemble the hardware and codify the software that will perform a number of functions, such as turning on and off power switches from your phone, detecting package deliveries and transmitting emails announcing their arrival, posting tweets on Twitter when your bird feeder needs to be refilled, and opening and closing curtains depending on light and temperature, and more.

Because the recommended skill set for building these solutions includes some familiarity with programming, this book builds upon several previously published Pragmatic Bookshelf titles. If you would like to learn more about programming Arduinos or writing Ruby or Python scripts, I strongly recommend checking out the books listed in Appendix 2, *Bibliography*, on page 211.

Each project begins with a general introduction and is followed by a What You Need section that lists the hardware parts used. This is followed by a section called Building the Solution that provides step-by-step instructions on assembling the hardware. *Programming Your Home* will call upon the Arduino extensively for most (but not all) of the projects. Once the hardware is constructed, it can be programmed to perform the automation task we built it to do. Programs can range from code for Arduino microcontrollers to scripts that execute on a computer designed to control, capture, and process the data from the assembled hardware elements.

The book concludes with a chapter on future projections in home automation and a chapter filled with idea starters that reuse the hardware and software approaches demonstrated in the eight projects.

Arduinos, Androids, and iPhones, Oh My!

With the meteoric rise of mobile device proliferation, the post-PC moniker has made its way into the tech world's vocabulary. I am a big proponent of technology shifts, but I am also old enough to have lived through three major computing revolutions (the shift from mainframes to PCs, the rise of the Internet, and the shift from PCs to mobile devices) and know that change isn't as fast as people say it is. Until mobile applications can be developed on mobile devices the way PC applications can be developed on PCs, a Linux, Windows, or Mac computer will be a central requirement for developing mobile apps. The same holds true for Arduino programming.

That said, the times are indeed a-changing. Microsoft Research was one of the first major phone OS providers to attempt to create native mobile applications directly on the mobile device with their release of TouchStudio. Google engineer Damon Kohler created the Scripting Layer for Android (SL4A) that gives Android users the ability to write fairly sophisticated programs using a text editor on their phone. Coupled with Sparkfun's IOIO ("yo-yo") board, we're already seeing early glimpses of what could replace the PC for some of the scripts created for this book.

Since you will need a Mac, Linux, or Windows computer to program the Arduinos and mobile apps in this book, this computer will also be the machine that runs the server-side programs that interpret and extend information out to your mobile devices. Of course, if you only have one computer and it's a laptop that travels with you, consider purchasing a cheap Linux or Mac to run as your home server. Not only will you benefit from having a dedicated system to run the monitoring apps 24/7/365, but it can also serve as your home Network Attached Storage (NAS) server as well.

I am a believer in open source hardware and software. As such, the projects in the book depend upon these. I am also technology-agnostic and rarely have any overriding devotion to one hardware supplier or programming language. Code for this book could have been presented just as easily in Mono-based C# and Perl, but I opted for Ruby and Python because of their portability and multiparty open source support. I could have used a Windows or Linux machine as the server and development system but chose Mac for the book because Ruby and Python are preinstalled with the OS, thereby eliminating the time and space required to install, configure, and troubleshoot the operating environment.

In accordance with this open source philosophy, I also opted to demonstrate the mobile application examples exclusively for the Android OS. While I personally prefer iOS devices as the platform of choice for my mobile lifestyle, the overhead associated with writing applications for iOS is a hassle. In addition to learning Objective-C and the various frameworks as well as dealing with the burden of memory management, deploying iOS apps requires either a jailbroken device or the legitimate purchase of an annual membership to Apple's iPhone developer network. Conversely, Android's SDK and application deployment is free and open. Android programs can also multitask better than iOS programs. Of course, these two advantages also bring on greater security and resource utilization risks. That said, I encourage readers who prefer the mobile demos to run on non-Android devices to port the simple client programs presented in this book to their favorite mobile OS and share these conversions with the Programming Your Home community.

Another term that is gaining a foothold in the tech press is the "Internet of Things." This phrase refers to the idea that with the proliferation of network-connected microcontrollers, Internet-based communication between such small devices will eventually outnumber people surfing the Web. While that may be the case for submitting data upstream, reaching such a device from the Internet is still a hassle. Besides the technical knowledge required to set up a dynamic DNS and securely configure port forwarding to easily reach the device, ISPs may block outbound ports to prevent end consumers from setting up dedicated servers on popular network ports like FTP, HTTP/S, and SMTP.

The projects in this book should work perfectly fine in a home local area network. However, obtaining sensor data outside of this local network is a challenge. How do you check on the status of something like a real-time temperature reading without going through the hassles of opening and forwarding ports on your router (not to mention the potential security risks that entails)?

Fortunately, several companies have begun to aggressively offer platforms accessible via simple web service APIs to help overcome these hassles. Three of these gaining momentum are Pachube, Exosite, and Yaler.[1] Configuring and consuming their services is a fairly straightforward process. I encourage you to visit these sites to learn more about how to incorporate their messaging capabilities into your own projects.

---

1. http://www.pachube.com, http://www.exosite.com, and http://www.yaler.org, respectively.

Code Examples and Conventions

The code in this book consists of C/C++ for Arduino, Java for Android, Ruby for web middleware, and Python for desktop scripts. Most of the code examples are listed in full, except when burdened by external library overhead (such as in the case of Android and Ruby on Rails program listings). Syntax for each of these languages is highlighted appropriately, and much of the code is commented inline along with bullet markings to help bring attention to the big ideas in the listings.

Highlights and sidebars are used sparingly in the book in an effort to keep information moving at a quick yet manageable clip.

Online Resources

Visit the book's website at http://pragprog.com/titles/mrhome, where you can download the code for all the projects, participate in the book's discussion forum, ask questions, and post your own home automation ideas. Bugs, typos, omissions, and other errors in the book can be found on the book's errata web page.

Other popular website resources include the popular DIY websites Makezine, and Instructables,[2] where participants share a wide variety of home-brewed creations with their peers.

There are also several IRC channels on freenode.net and SIG forums on Google Groups dedicated to the subject, with many focused on singular aspects of DIY gadget design, home automation, and hardware hacking.[3]

OK, enough with the preamble. Let's get ready to build something!

---

2. http://www.makezine.com and http://www.instructables.com, respectively.
3. http://groups.google.com/group/comp.home.automation/topics

Part I

Preparations

CHAPTER 1

Getting Started

Before we start wiring up hardware and tapping out code, let's lay down the foundation, starting with what exactly we mean by home automation, what's been available in the consumer space in the past, and why building our own solutions makes sense today and in the future.

We will also review a couple of design and construction best practices that will be put to good use when assembling the projects in this book.

We'll start by defining what we mean by home automation. Next we'll consider some of the prepackaged commercial solutions on the market, and then we'll take a quick snapshot of some of the more popular custom automation hardware and software projects. The chapter will conclude with some of the tools and practices that have helped me quite a bit when building the projects in this book as well as with other projects beyond the home automation category.

1.1 What Is Home Automation?

So what exactly does the term home automation mean? At its most basic level, it's a product or service that brings some level of action or message to the home environment, an event that was generated without the homeowner's direct intervention. An alarm clock is a home automation device. So is a smoke alarm. The problem is, these stand-alone devices don't use a standard network communication protocol, so they can't talk to one another the way that networked computers can.

One of my earliest memories of home automation was when the Mr. Coffee automatic drip coffee machine came out in the early 1970s. The joy this simple kitchen appliance brought my coffee-drinking parents was genuine. They were so pleased to know that when they woke up in the morning a freshly brewed pot of coffee would be waiting for them. Who would have thought that such a simple concept as a coffee maker combined with an alarm clock would change their world?

Now that we're in the twenty-first century, rudimentary coffee makers are getting a makeover by tinkerers bolting network adapters, temperature sensors, and microcontrollers to make the brew at the right time and temperature and to send a text message alert that the beverage is ready for consumption. It's only a matter of time before manufacturers incorporate inexpensive electronics into their appliances that do what tinkerers have been doing with their home electronics for years. But a standard communication protocol among such devices remains elusive. Nevertheless, efforts are afoot by a number of home automation vendors to address that problem.

1.2 Commercial Solutions

The number of attempts to standardize home automation communication protocols has been ongoing nearly as long as Mr. Coffee has been in existence. One of the earliest major players was X10, a company that still offers basic and relatively inexpensive home automation solutions today. X10 takes advantage of existing electrical wiring in the home. It uses a simple pulse code protocol to transmit messages from the X10 base station or from a computer connected to an X10 communication interface. But problems with signal degradation, checksums, and return acknowledgments of messages, as well as X10's bulky hardware and its focus on controlling electrical current via on/off relay switches, have constrained X10's broader appeal.

Other residentially oriented attempts at standards, such as CEBus and Insteon, have been made, but none have attained broad adoption in the home. This is partly due to the chicken-and-egg problem of having appliance and home electronics manufacturers create devices with these interfaces and protocols designed into their products.

Most recently, Google has placed its bet on the Android operating system being embedded into smart devices throughout the home. Time will tell if Google will succeed where others have failed, but history is betting against it.

Rather than wait another twenty years for a winning standard to emerge, embedded computing devices exist today that employ standard TCP/IP to communicate with other computers. This hardware continues to drop to fractions of the prices they cost only a few years ago. So while the market continues to further commoditize these components, the time is now for software developers, home automation enthusiasts, and tinkerers to design and implement their own solutions. The lucky few will uncover and market a cost-effective, compelling solution that will one day catch on like wildfire and finally provide the impetus to forever change our domestic lives.

1.3 DIY Solutions

The Do-It-Yourself category in home automation is more active today than ever before. The combination of inexpensive electronics with low-cost networked computers make this option extremely attractive. There's other reasons that make DIY an ideal pursuit. Unlike proprietary commercial offerings, the projects you build are not mysterious black boxes. You have the source code. You have the knowledge. You have the measurements, the metrics, and the methods.

Not only will you know how to build it, you will know how to troubleshoot, repair, and enhance. None of the commercial solutions can match exactly what you may need. Home automation vendors have to generalize their products to make them appeal to a large consumer base. By doing so, they don't have the luxury of creating one-off solutions that exactly match one customer's specific needs. But with some rudimentary knowledge and project construction experience, you'll gain the confidence to create whatever design matches your situation.

For example, the first project in this book builds a sump pit notifier that emails you when water levels exceed a certain threshold. While commercial systems have audible alarms, none that I have found at the local hardware store have the means to contact you via such messaging. And should you need to modify the design (add a bright flashing LED to visually broadcast the alert, for example), you don't need to purchase a whole new commercial product that includes this feature.

Walk around your house. Look for inefficiencies and repetitive tasks that drive you crazy the way George Bailey was with pulling off the loose finial on his staircase's newel post. Take note of what can be improved with a little ingenuity and automation. You may be surprised at just how many ideas you can quickly come up with.

1.4 Justifying the Investment

Let's be honest. Spending more money on parts that may or may not work well together versus buying a cheaper purpose-built device that meets or exceeds the functionality of a homegrown solution is simply not a good investment. Sure, there may be some value derived from the knowledge gained from the design experience, the pleasure of building the solution, and the satisfaction of seeing your creation come to life. But justifying such an investment to a budget-conscious spouse, for example, may deflate whatever gains you have made in the satisfaction department.

When considering any new design approach, strive for a scenario where you will maximize your time, equipment investment, and learning potential. You may have to try several experiments and iterations before the hardware and software come together and work the way you envisioned. But if you keep at it, you will be well rewarded for your persistence. Not only will you achieve high points for devising a low-cost solution, but such constraints will help drive creativity to even higher levels. That's why I have tried my best to keep all the projects in this book within a reasonable budget, and I encourage reuse of old electronic parts and materials as much as possible.

Do your homework. Research online to see who may have attempted to build what you have in mind. Did they succeed? Was it worth the money and time they invested? Is there a commercially viable alternative?

If you determine that your idea is unique, put together an estimate of the expenses in terms of your time and of the materials you need to purchase. Remember to also include the cost of any tools you need to buy to construct and test the project's final assembly. This added expense is not negligible, especially if you're just starting down the DIY road. As you get more involved with hardware projects, you will quickly find that your needs will expand from an inexpensive soldering iron and strands of wire to a good quality multimeter and perhaps even an oscilloscope. But the nice thing about building your own solutions is that you can build them at your own pace. You will also find that as your network of DIYers grows, your opportunities for group discussion, equipment loans, insightful recommendations, and encouragement will grow exponentially.

1.5 Setting Up Your Workbench

Good assembly follows good design. Building these projects in a frustration-free environment will help keep your procedures and your sanity in check.

Work in a well-lit, well-ventilated area. This is especially important when soldering. Open a window and use a small fan to push the fumes outside. Use a soldering exhaust fan if an open window isn't an option.

If your work space can afford it, have a large table to spread out your electronic parts. Keep it close to power outlets and have a power strip on the table for easy access.

Organize your components with small craft containers, baby food jars, pill boxes, Altoids tins—anything that helps keep the variety of capacitors, resistors, LEDs, wires, shields, motors, and sensors sorted will make it much easier to keep track of your parts inventory.

Have your computer stationed near or on the work space. This is a no-brainer if it's a laptop. If it's a desktop, minimize its table footprint by only placing a monitor, mouse, and keyboard (both preferably wireless) on the table to leave as much unobstructed working space as possible.

Keep clutter away from underneath and around the table. Not only does this aid fire prevention, but doing so will also make it far easier to find that elusive component when it rolls off the table and bounces toward the unknown.

Lastly, keep the work space dedicated to project work. Some projects can be like building a jigsaw puzzle. You need a place for the half-assembled pieces to sit while life goes on. Being able to sit down and start working, rather than start unboxing and repackaging a fur ball of wires and parts, makes building projects a joy instead of a chore.

1.6 Sketching Out Your Ideas

When inspiration strikes, nothing beats old-fashioned pencil and paper to quickly draw out your ideas. For those who prefer to brainstorm their designs on a computer, several free, open source, cross-platform tools have helped me assemble my ideas and document my work:

- Freemind is great for organizing thoughts, objectives, and dependencies.[1] This mature mind-mapping application helps you make sense of a brain dump of ideas and see the links between them. This will save you time and money because you will be able to spot key ideas, eliminate redundancies, and prioritize what you want to accomplish.

- Fritzing is a diagraming application specifically designed for documenting Arduino-centric wiring.[2] Unfortunately, it's still a work in progress and is rough around the edges. It also doesn't have a number of the popular sensors iconically represented yet, but the object library is growing as

---

1. http://freemind.sourceforge.net
2. http://fritzing.org/

more people contribute to the project. I use this application exclusively for documenting my Arduino-based projects, which is why the wiring diagrams in this book were generated by Fritzing.

- Inkscape is an easy-to-use vector-based drawing program that helps sketch out ideas beyond the Arduino-centricity of Fritzing.[3] While Inkscape is mainly intended for graphic artists, it has accurate measurement tools that are great for scoping out bracket and enclosure ideas for your projects.

Going beyond the desktop, tablets are rapidly taking over the role that were once the domain of traditional paper uses. In fact, it wouldn't surprise me if you're reading this book on an iPad or a Kindle right now. Beyond just reference lookups, tablets are excellent for brainstorming ideas and creating initial sketches of preliminary project designs. An iPad (or Android tablet, for that matter) combined with a sturdy stand also makes for a handy electronic reference. Load up your sketches, track your progress, reorder priorities, and make notes along the way.

My current favorite iPad apps for my projects include the following:

- Elektor Electronic Toolbox is an electronic parts reference with a variety of helpful calculators and conversion tools.[4]
- iCircuit is a electronic circuit simulator that makes building and understanding circuits far easier than static diagrams on a printed page.[5]
- iThoughts HD is a mind-mapping application compatible with importing and exporting Freemind files.[6]
- miniDraw is a vector-based drawing program that can export to SVG format, perfect for importing your sketches into Inkscape.[7]

In addition to designing and documenting your projects, well-executed projects also rely on taking accurate measurements and running tests to validate your work.

1.7 Writing, Wiring, and Testing

Unfortunately, no good software emulator exists yet for the Arduino; fortunately, programs for this platform are usually small and specific enough such

---

3. http://inkscape.org
4. http://www.creating-your-app.de/electronic_toolbox_features.html?&L=1
5. http://icircuitapp.com/
6. http://ithoughts.co.uk
7. http://minidraw.net/

that the compile-run-debug cycles are tolerable. Good coding and testing techniques go a long way toward ensuring a high-quality outcome. The same goes for constructing and wiring up the physical electrical connections.

While nearly all of the projects in this book can be constructed without solder, permanent installations require good soldering techniques to ensure a conductive pathway. It's best to verify (usually with the help of a breadboard) that the connections work as expected before making them permanent with solder.

Use good code-testing techniques. Whether for microcontroller code for the Arduino or server-side code for your Ruby or Python scripts, Test-Driven Development (TDD) is a good practice to adopt. There are a number of good testing frameworks and books available on the subject. Read Ian Dees's article, "Testing Arduino Code," in the April 2011 edition of *PragPub* magazine,[8] as well as *Continuous Testing: with Ruby, Rails, and JavaScript* [RC11].

Run unit tests like py.test when writing Python-powered scripts. When coding in Ruby and creating Rails-based web front ends, consider using Rspec (for more details on using Rspec, read *The RSpec Book* [CADH09]). Use the Android testing framework for your Android applications.[9] Even when working on small applications, using proven testing methodologies will help keep you sane while further elevating the quality in your code.

Know how to use a multimeter. Like a software debugger, a multimeter can come in quite handy when trying to figure out what's happening inside your project—for example, where a short might be stepping on your project. Besides detecting problems, a multimeter is also useful for measuring electrical output. For example, you can also use it to determine if a solar battery pack can deliver enough uninterrupted energy to power a microcontroller-operated servo.

If you're not familiar with how a multimeter operates, just type "voltmeter tutorial video" in your favorite search engine. There are plenty online to choose from.

1.8 Documenting Your Work

Hand-drawn scribbles offer nice starting points, but often projects take twists and turns along the way that have to account for limited resources or hardware

---

8. http://www.pragprog.com/magazines/2011-04/testing-arduino-code
9. http://developer.android.com/guide/topics/testing/testing_android.html

that just doesn't work as planned. The final product may be vastly different from the original design. That's why it's so important to finish a project with accurate, clean, and concise documentation, especially if you plan to share your design with others.

Using applications like Fritzing can aid with the generation of clean, full-color wiring diagrams. Doing so will go a long way toward showing exactly how to wire up a project. Nothing is worse than seeing blurry, angled Flickr photos or YouTube videos of wires plugging into hard-to-see shield or breadboard pinholes as the primary means of documentation. Having those are nice supplementals, but any well-designed project should be accompanied by clear and easy-to-follow wiring illustrations.

Leave verbose comments in your code, even for the simple scripts and sketches. Not only will it help you and those you share the code with understand what various routines are doing, good comments will also put you back in the frame of mind you were in when writing the code in the first place. And if you share your code on various repository sites like Github and Sourceforge, well-commented code shows a greater level of professional polish that will gain you more respect among your peers.

With all these recommendations, keep in mind that the most important takeaway from the book's projects is to have fun doing them. These rewarding experiences will encourage you to use these projects as starting points and infuse your own unique needs and design goals into them.

In the next chapter, we will review the hardware and software we will use and take into account the optimal configurations of each.

CHAPTER 2

Requirements

Before diving into the book's projects, we need to consider the materials and best practice methodologies we will employ when building the solutions.

A key tenet I practice in this book is for the various projects to be as easy and inexpensive to build as possible. While it may be fun to construct an elaborate Rube Goldberg contraption that costs hundreds of dollars to open a can of soup, it's far more practical to spend a dollar on a can opener that you can buy from the store. I have tried my best to maximize the value of money and time with each project. As such, few of them should cost more than sixty dollars in parts or take more than an hour to construct.

It's also good to practice reuse whenever possible. This is far easier for software than for hardware, but it can be done. That is why an inexpensive microcontroller board like the Arduino is at the center of several of these projects.[1] In an effort to save money on the hardware investment, it may be worthwhile to try out one or two projects concurrently and decide which ones make the most positive impact before buying a half dozen Arduino boards. After you have built the projects that you're most interested in, then build upon them, improve them, and remix them. When you have an especially cool creation, contribute your discoveries to the Programming Your Home book forum.

Most software development projects typically do not require much more than a computer and the choice of language and frameworks the programming logic executes within. But with the addition of hardware sensors, motors, and purpose-built radios and controllers, the design and construction workflow is a little more complex. Essentially, you are building two major components with each project: the physical collection of hardware and the software that

---

1.   http://arduino.cc/

will measure, interpret, and act on the data that the hardware collects. Let's take a look at what comprises these two key development aspects.

2.1 Knowing the Hardware

The Arduinos, sensors, and motors (technically referred to as actuators) used in the projects can be purchased from a number of online retailers, with my current two favorites being Adafruit Industries and Sparkfun.[2] For the budget-conscious builder, Craigslist and eBay offer money-saving deals. Purchasing used parts from these online classified listing services may come in especially handy when searching for old Android handsets and X10 controls. But buyer beware: there is often little recourse you can take should a used part stop working a few days after you have received it. Companies like Adafruit and Sparkfun stake their reputations on their over-the-top customer support and will usually accommodate any reasonable replacement request.

Each project in the book contains a What You Need section that lists the hardware and software components required to build the solution. The hardware used is nothing exotic or difficult to find and purchase online, and some projects even incorporate common household items like dry cleaning clothes hangers and cloth scraps in their parts list. Here is a complete inventory of electronic components required to build the projects in this book and their estimated per item costs:

- Arduino Uno, Duemilanove, or Diecimila - $30
- Ethernet shield - $45
- Wave shield with speaker, wire, and SD card - $35
- Passive infrared (PIR) motion sensor - $10
- Flex sensor - $12
- Force sensitive resistor - $7
- TMP36 analog temperature sensor - $2
- CdS photoresistor (commonly referred to as a photocell) - $1
- Standard servo motor - $15
- Smarthome electric 12VDC door strike - $35
- Two XBee modules and adapter kits - $70

---

2. http://www.adafruit.com and http://www.sparkfun.com, respectively.

- FTDI connector cable - $20
- Solar charger with built-in rechargeable battery - $30
- X10 CM11A ActiveHome serial computer interface - $50
- X10 PLW01 standard wall switch - $10
- Serial USB converter - $20
- Home computer (Linux or Mac preferred) - $200 to $2,000, depending on model
- Wireless Bluetooth speaker - $120
- Android G1 phone - $80 to $150, depending on its used condition
- Android smartphone - $50 to $200, depending on features and carrier contract
- Sparkfun IOIO board with JST connector, barrel jack to 2-pin JST connector, and 5VDC power supply - $60
- Male USB to male mini-USB cable - $3
- 2.1 mm female barrel jack cable - $3
- Spool of wire (22 AWG should be adequate) - $3
- 10K ohm resistor - $0.10
- 10M ohm resistor - $0.10
- Small breadboard - $4
- Electrical tape or heat shrink tubing - $5
- 9-volt DC power supply - $7
- 12-volt 5A switching power supply - $25
- PowerSwitch Tail II with a 1K resistor and a 4222A NPN transistor - $20
- Stepper motor - $14

Each of these parts is reusable with the projects throughout the book. Naturally, if a particular project is permanently installed in your home, you will have to replenish the inventory to replace the parts used in that permanent fixture. Do It Yourself (DIY) hardware project building, like writing code, is a satisfyingly addictive experience. As your confidence grows, so too will your expenditures on electrical components.

Of all the parts used throughout the book, three items that are frequently called upon are Android smartphones, Arduinos, and XBee radios. I will give a brief overview of each in the next sections. If you intend to leverage these useful electronics further, refer to the Android, Arduino, and XBee titles in Appendix 2, *Bibliography*, on page 211, for more information on these remarkable, transformative technologies.

Android Programming

The Android operating system is continuing its rapid expansion and domination in certain telecommunications and embedded systems markets. Google announced its Android@Home initiative and is encouraging developers and consumer electronics manufacturers to consider Android as a base technology for smart home systems. Several electronics vendors have released hardware that is compliant with the Android Open Accessory Development Kit (ADK) and that takes advantage of the interfaces Google has designed.[3]

The ADK board I chose is Sparkfun's IOIO board. ADK support for the IOIO was still in beta at the time of this book's publication, and loading the ADK-enabled firmware on the board is not a trivial exercise. Chapter 9, *Android Door Lock*, on page 141, instead discusses a project in this book using traditional Android SDK calls while incorporating the custom hardware library that the IOIO board currently provides.

As the cost of ADK developer hardware drops, more economically viable options will be available for developers and manufacturers alike. But for now, a used first-generation Android phone coupled with an IOIO is still far more powerful and much less expensive than a comparably spec'd ADK board with the same features (camera, GPS, Bluetooth, Wi-Fi) as a smartphone. By the time ADK devices become cheap and plentiful, you will be ahead of the game by having working knowledge of the Android application development ecosystem.

Some Android-centric projects involve building both a native client and a server application. While the client applications could have been written in a device-agnostic web framework like jQuery Mobile,[4] it's useful to stress the importance of native mobile app development. By having this native foundation from the start, you will be able to more easily call upon advanced phone functions that are inaccessible from a web-based interface. Native applications also tend to load and respond faster than their browser-based counterparts.

---

3. http://developer.android.com/guide/topics/usb/adk.html
4. http://jquerymobile.com/

While it's not necessary to have prior experience developing Android applications to build the Android programs in this book, it will certainly help to have some familiarity with the Android SDK.[5]

Arduino Programming

If you have C or C++ coding experience, you will feel right at home with writing code for the Arduino's ATMega 168/328 microcontroller. Arduino programs, known as *sketches*, are easy to write once you learn the basic structure of an Arduino application.

Let's take a quick look at the basic anatomy of an Arduino sketch. It begins with #include statements at the head of the sketch import code libraries, just as they are in C programs. This is followed by global variable and object initializations that are usually referenced in the sketch's setup() routine. The setup() function is typically used to reference physical wiring connection points, known as "pins" on the Arduino board, along with the global variable assignments made in the initialization section. An example of this assignment might be something like int onboard_led = 13; before setup(). This code instructs the Arduino to use pin 13 (the location of its onboard LED) to be accessible in the sketch. We can then assign the pin for output with the line pinMode(onboard_led, OUTPUT) within the setup() routine.

After the variable assignment and setup() program initializations are established, sketches enter the main loop() routine that infinitely iterates over the instructions contained within it. It is here that the sketch waits for some event to occur or repeats a defined action. We will revisit this structure and the process of writing, compiling, and running Arduino programs again in our first project, the *Water Level Notifier*.

Any text editing program can be used to write sketches, with the most popular being the free Arduino Integrated Development Environment (IDE) available for download from the Arduino website. This Java-based coding environment incorporates everything you need to compile your sketches into machine-friendly ATMega microcontroller instructions. It also comes bundled with dozens of sample sketches to help you quickly learn the syntax and realize the number of different sensors and motors that the Arduino can interact with. And because it is based on Java, the Arduino IDE will run identically on Windows, Mac, and Linux computers.

---

5. http://developer.android.com/sdk

> **Joe asks:**
> ### Does the Arduino IDE Have a Virtual Emulator?
>
> Unlike most desktop and mobile application development, no official Arduino emulator exists. It's difficult to simulate all the different physical sensors and motors that the Arduino can be connected to. Several third-party attempts have been made to create such a tool, but they are either limited in the operating systems they support or focus on the ATMega chip and not the full Arduino package. Two Windows-based emulators are Virtual Breadboard and Emulare,[a] with Virtual Breadboard being the one I recommend due to its virtual representation of Arduino hardware. Virtual Breadboard also provides a limited set of emulated sensors and other devices that connect to the onscreen Arduino.
>
> Given the low cost of the Arduino itself, few find much use for an emulator other than for unit testing and convenient, portable virtual hardware reasons. Spend the money for an actual board rather than messing around with the emulators. Sketches are short, and the serial window in the Arduino IDE is helpful enough to offer detail to adequately debug and tweak real-live hardware.
>
> ---
>
> a. http://www.virtualbreadboard.net and http://emulare.sourceforge.net/, respectively.

XBee Programming

Another key technology we will be using in several of the projects is a radio device based on the IEEE 802.15.4 wireless specification, commonly known as XBee. XBee radios are ideal for Arduino-based wireless projects due to their low-cost, low-power, and easy-to-use serial interface communication. Low-powered XBees are used mainly for character-level bitstream communications. Broadcast distances between radios are roughly within a fifteen-meter (50-foot) radius.

The projects in this book that incorporate XBee-to-XBee communications use single characters or short strings to announce a state change as a result of a sensor event. Such changes are then broadcast wirelessly to a paired XBee modem that is usually attached to a computer or embedded system that processes the received signal. I prefer to log this data before acting upon it to store events and help with debugging. After logging the received data, the computer may also further propagate the signal by translating it into a web service-friendly payload, an email message, a servo motor movement, or any other call to action.

The most time-consuming and challenging aspect of using XBees is correctly assembling the hardware and pairing the radios. It is not a trivial procedure, but it is also not too difficult either. Fortunately, Limor "Ladyada" Fried, founder of Adafruit Industries and open hardware electrical engineer extraordinaire, has posted a terrifically helpful tutorial on her website that provides detailed, step-by-step instructions on assembling XBee adapter kits sold along with the XBee radio modules. We will explore this further when we use XBees for the first time in the *Tweeting Bird Feeder* project.

Incidentally, Digi International, the company who manufacturers the XBee hardware, recently announced a 802.11 b/g/n Wi-Fi–capable XBee that obviates the need for a second XBee connected via an FTDI cable for the receiving PC. However, the cost for this convenience is considerably more than the configuration I used in the book. If you're interested in this more convenient approach, check out the XBee Wi-Fi page on Digi's website.[7]

A number of books (such as *Building Wireless Sensor Networks* [Fal10]) and online resources go into greater detail on learning basic electronics, Arduino programming, and wireless networking. This section simply provided an overview of how to work with the specific hardware we will use in this book's projects. In the next section, we will take a quick survey of the software we will use to bring the assembled hardware to life.

2.2 Knowing the Software

In addition to being familiar with the C/C++ syntax used for programming Arduino sketches, you will be able to follow along easier if you are familiar with the Java, Ruby, and Python languages. Ruby on Rails experience is also a plus. If you are unfamiliar with these, review Appendix 2, *Bibliography*, on page 211, for several titles that do a great job of teaching these languages and frameworks.

Even if you don't know much about these languages, you should be able to build and execute the code for these projects with little or no modification on a Linux or Macintosh computer. Windows users will need to install their preferred Python and Ruby distributions as well as the Java runtime, and some of the utilities used in this book that were written for Unix-based operating systems might not have a Windows version available. A PC can be loaded with your preferred Linux distribution, and a Mac Mini will be more than adequate for the OS X crowd. This home server should be a reasonably inexpensive component in the Programming Your Home hardware collection.

Java familiarity will come in handy when writing the Android client and server applications later in the book. Experience with Python and Ruby is

---

7. http://www.digi.com/xbeewifi

also a plus. Python also comes preinstalled on Mac and on nearly all Linux distributions. As such, a majority of server-side scripts in this book are Python-based. Java, Perl, PHP, or Ruby developers intent on staying pure to their favorite technology shouldn't have too difficult a time porting the project's server-side applications to their language of choice. I encourage any readers interested in porting the book's code to a different language to share their work with other readers via the book's website.

2.3 Be Safe, Have Fun!

I deliberately designed the projects in this book to have little or no chance of electrical shock or damage to persons or property should something unexpected occur. It should go without saying that you should always employ safe practices when assembling any hardware project.

In addition to the book's disclaimers that I as the author and The Pragmatic Bookshelf as the publisher cannot be held liable for any damages of any kind as a consequence of building and powering these projects (as well as be held liable in any way for hardware you use or modify for these projects—for specific details, see *Proceed at Your Own Risk. You Have Been Warned!*, on page 19), I cannot stress this highly enough: unless you are a certified electrician, plumber, or carpenter and know exactly what you're doing at all times, don't start poking around and tampering with the basic infrastructures found in the home. Call upon the experience of professional, certified electricians when wiring for the home. Trust me. The up-front planning and outside expertise will deter aggravation, save you money, and protect you from physical harm. Leaving these foundational aspects to the professionals will leave you with more time to implement and optimize your ultra-cool and envy-invoking smart home creations.

OK, enough with the requirements and disclaimers. Let's dive into the next section, where we will finally get to assemble and code some really nifty and unique home automation projects!

Proceed at Your Own Risk. You Have Been Warned!

Your safety is your own responsibility. Use of the instructions and suggestions in this book is entirely at your own risk. The author and the Pragmatic Programmers, LLC, disclaim all responsibility and liability for any resulting damage, injury, or expense as a result of your use or misuse of this information.

It is your responsibility to make sure that your activities comply with all applicable laws, regulations, and licenses. The laws and limitations imposed by manufacturers and content owners are constantly changing, as are products and technology. As a result, some of the projects detailed here may not work as described or may be inconsistent with current laws, regulations, licenses, or user agreements, and they may even damage or adversely affect equipment or other property.

Power tools, electricity, and other resources used for these projects are dangerous unless used properly and with adequate precautions, including proper safety gear (note that not all photos or descriptions depict proper safety precautions, equipment, or methods of use.) You need to know how to use such tools correctly and safely. It is your responsibility to determine whether you have adequate skill and experience to attempt any of the projects described or suggested here. These projects are not intended for use by children.

Make sure you are comfortable with any risks associated with a project before starting that project. For example, if the idea of dealing with 110V power worries you, then don't do the projects that use it, and so on. We also don't know about any local ordinances that might apply to you, so before you go wiring stuff in, you should check your building codes. If in doubt, have a chat with a local professional.

Only build these projects if you agree that you do so at your own risk.

Good luck, and have fun!

Part II

Projects

CHAPTER 3

Water Level Notifier

If you live in the midwestern part of the United States like I do, you know all about heavy rains and the effects they can have on a basement. Stories of sump pumps failing during torrential downpours are often punctuated with "Had I only known how quickly the water level in my sump pit was rising, I would have had more time to move my stored items out of the way."

Imagine another scenario, where someone needs to use a dehumidifier to remove dampness in a cellar. Inexpensive dehumidifiers often stop working when water reaches a certain height in the condensation bucket. Some models may include an audible alarm or flashing light when this shutdown occurs, but such alerts are ineffective because the dehumidifier is typically installed in an infrequently visited area.

Wouldn't it be more convenient to receive an email from your house when the water levels in these containment areas exceed a certain threshold, alerting you to take action? (See Figure 1, *Have your house email you*, on page 24.) Let's get our feet wet, so to speak, and build a system that will provide this helpful notification service.

3.1 What You Need

The main component required to make this project work is something called a flex sensor. The buoyancy of rising water levels will bend the sensor. As the sensor bends one way or the other, current values will increase or decrease accordingly. The sensor's position can be read with a simple Arduino program and can be powered via either the Arduino's 3.3 or 5.0 volt pins.

Here's the complete list (refer to the photo in Figure 2, *Water Level Notifier parts*, on page 25):

Chapter 3. Water Level Notifier

Figure 1—Have your house email you ...when water levels rise precipitously to give you enough time to prevent flood damage. This project can also be used to monitor water levels in dehumidifiers, air conditioners, and similar devices.

1. An Arduino Uno
2. An Ethernet shield[1]
3. A flex sensor[2]
4. A 10k ohm resistor[3]
5. A 1-inch fishing bobber

---

1. http://www.adafruit.com/index.php?main_page=product_info&cPath=17_21&products_id=201
2. http://www.sparkfun.com/products/8606
3. http://www.makershed.com/ProductDetails.asp?ProductCode=JM691104

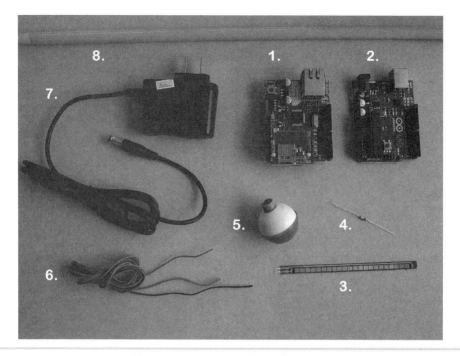

Figure 2—Water Level Notifier parts

6. Three wires (power, ground, and analog pin 0) trimmed to desired length
7. A 9-volt power supply to power the Arduino and Ethernet shield once untethered from the USB cable
8. A pole or wood plank to attach and hang the flex resistor from
9. A web server running PHP 4.3 or higher (not pictured)

You will also need a standard A-B USB cable (not pictured) to connect the Arduino to the computer and an Ethernet cable (also not pictured) to connect the Ethernet shield to your network.

We will be reusing the Arduino and Ethernet shield again in several other projects, so—not including the cost of these two items—the remaining hardware expenses should be under twenty dollars. Considering the peace of mind and the ease with which you can build further ideas upon this concept, this is money well spent.

3.2 Building the Solution

Before the Water Level Notifier can start broadcasting alerts, we need to complete the following tasks:

> **Arduino Ethernet**
>
> Would you prefer a board that combines the Arduino Uno and the Arduino Ethernet shield into a single package? The Arduino Uno Ethernet may be what you're looking for.[a] However, the board still needs to reserve digital pins 10 through 13 for the Ethernet module, just like the separate Ethernet shield does. The Arduino Uno Ethernet also requires an FTDI cable to interface with a computer rather than the more popular A-B USB cable.[b] The biggest advantage that this board has to offer is the ability to combine Ethernet services with another Arduino shield, assuming that shield does not require the same pin resources that the Ethernet hardware requires.
>
> ---
>
> a. http://www.adafruit.com/products/418
> b. https://www.adafruit.com/products/70

1. Attach wires and a resistor to the exposed sensor leads on one end of the flex resistor and the fishing bobber on its other end.

2. Connect the leads of the flex sensor to an analog pin of an Arduino.

3. Write a program (i.e., sketch) for the Arduino that will monitor changes in the flex sensor readings. It should trigger an event when a large-enough deviation from the initial value is detected.

4. Attach an Ethernet shield to the Arduino so that the sketch can communicate with a web server running PHP.

5. Write a PHP script that will capture incoming values from the Arduino. When the water level has changed, it should format a message and send an email alert to the intended recipient, who will need to react quickly to the alert!

We will begin by assembling the hardware and testing out the flex sensor measurements.

3.3 Hooking It Up

Let's start by making sure our flex sensor works the way we intend it to. Connect the positive lead of the sensor to the Arduino's 5.0-volt pin using a wire. When looking at the flex sensor standing on its end, the positive lead is the trace that runs vertically. The negative lead is the one that looks like the rungs of a ladder. Connect the negative lead to the analog 0 pin with another wire. Lastly, bridge the analog 0 pin to the ground pin using the 10k ohm resistor to dampen the flow of current through the circuit. Refer to Figure

3, *Water Level Notifier wiring diagram*, on page 28, to make sure you attach the wires and resistor to the correct pins.

Attach the bobber to the end of the flex sensor. Most bobbers come with a retractable hook that can be fastened to the plastic tip of the sensor. If the bobber doesn't stay affixed to the sensor, you can also use hot glue or heat shrink tubing to help keep the bobber attached. Just be careful not to damage the sensor when heating it with these affixing solutions. You can also try duct tape as a safe alternative, though the tape may lose its grip over time.

Use plenty of wire so you have enough length to safely mount the Arduino and power source far away from the water source. The Arduino that I have monitoring my sump pit is sitting in a hobby box mounted on the wall several feel above the sump pit, and the two wires attached to the flex resistor are about two meters (roughly six feet) in length.

Now that the Arduino has been wired up, we can work on the logic of what the hardware is supposed to do for us. We will begin with a quick test program that will verify that the flex sensor is connected correctly and working properly.

3.4 Sketching Things Out

Before we start writing code, we first need to make sure we can communicate with the Arduino. Then we will learn how to collect and act upon data sent by the flex sensor with a program (what the Arduino community prefers to call a *sketch*).

The first sketch we write will detect when the flex resistor values have changed. If the change is large enough (in other words, if water is making the resistor bend), we will transmit a request to a PHP server that will process the request. That server will then send out an email notifying us of the change.

We will build the sketch incrementally, first by connecting the flex sensor to the Arduino and collecting values when the sensor is straight and then when it bends in both directions. Once these values have been identified, we will write conditional statements that will call functions to send HTTP GET statements containing data we will include in the email alert.

Configuring an Arduino

We will use the Arduino IDE to write, compile, and download our code into the Arduino. For those who would like a more comprehensive introduction to Arduino programming, read Maik Schmidt's excellent *Arduino: A Quick Start Guide* [Sch11].

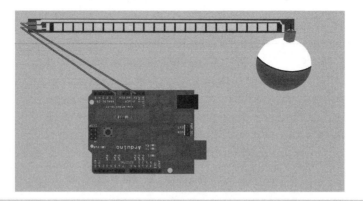

Figure 3—Water Level Notifier wiring diagram

If you are already familiar with the Arduino or are willing to hang on for the ride, let's get started by launching the Arduino IDE. Check to ensure that your Arduino is connected via USB cable and recognized and selected accordingly on one of the serial ports identified by the Arduino IDE's Tools→Serial Port menu. You can perform a quick test of your configuration with the LED Blink example program located in the Arduino IDE's File→Examples→1.Basics →Blink menu. Upload it to the attached Arduino and check to see that it executes correctly.

If it fails to do so, first verify that the Arduino is correctly plugged into the computer and powered by the USB. If it is, check next to be sure you've selected the correct serial port in the Arduino IDE and highlighted the right type of Arduino board in the Tools→Board. A few correctly placed mouse clicks on either of these settings usually fixes the problem.

The Flex Sensor Sketch

Now that the Arduino is connected and tested, we can write a sketch that will validate and interpret the bending of the flex sensor. We will begin by defining a few constants that we will refer to in the program.

Since we have to account for the sensor bending in either direction, we will define two named constants that will be used to set the upper and lower threshold limits.

We place these constants at the beginning of the sketch so they're easier to locate in case we need to edit these values later on. By convention, defined constants are also all uppercase so that they are easier to identify in the code. Let's call them `FLEX_TOO_HI` and `FLEX_TOO_LOW`. The range between these upper and lower limits will depend on the degree of flex that is optimal for your own scenario. I prefer a variance of plus or minus five units to allow a minor amount of bend before the notification event is triggered. Having such a range will allow us to account for minor environmental effects like a light breeze or a low-grade vibration.

We also need to account for the Arduino's onboard LED and the analog pin that the flex sensor is attached to.

- FLEX_TOO_HIGH is the value of the assigned analog pin when the flex sensor is bent forward past this threshold.

- FLEX_TOO_LOW is the value of the assigned analog pin when the flex sensor is bent backward past this threshold.

- ONBOARD_LED is assigned to the Arduino's onboard LED located at pin 13. We will use it provide us with a visual indicator when the flex resistor has deviated far enough to send an alert. This allows us to use the Arduino's onboard LED as a kind of visual debugger so that we can visually confirm that the flex events are being detected.

- FLEX_SENSOR is connected to the analog pin on the Arduino that the flex resistor is connected to. In this case, that value is 0 because the resistor is connected to pin 0.

These constants will be defined at the beginning of the sketch.

WaterLevelNotifier/WaterLevelSensor.pde
```
#define FLEX_TOO_HI    475
#define FLEX_TOO_LOW   465
#define ONBOARD_LED    13
#define FLEX_SENSOR    0
```

Now we will create two variables to capture the changing value and state of the flex resistor and set their initial values to zero.

- bend_value will store the changing analog values of the flex resistor as it bends.

- bend_state is the binary condition of the flex sensor. If it's straight, its value is equal to zero. If the flex resistor deviates either direction, we will set its state to one.

These variables will follow after the define statements we wrote earlier.

WaterLevelNotifier/WaterLevelSensor.pde
```
int  bend_value = 0;
byte bend_state = 0;
```

With the constants defined and the variables initialized, we need to set up the serial port to monitor the continuous stream of values being polled in the main program's loop. The onboard LED also has to be configured so we can see it turn on and off based on the bend_state of the flex resistor.

WaterLevelNotifier/WaterLevelSensor.pde
```
void setup()
{
    // for serial window debugging
    Serial.begin(9600);
    // set pin for onboard led
    pinMode(ONBOARD_LED, OUTPUT);
}
```

With the upper and lower flex bending limits defined, we need a routine that will check to see if these limits have been exceeded. If they have, we will turn on the Arduino's onboard LED. When the flex resistor returns to its resting straight position, we will turn the LED off.

WaterLevelNotifier/WaterLevelSensor.pde
```
void SendWaterAlert(int bend_value, int bend_state)
{
    digitalWrite(ONBOARD_LED, bend_state ? HIGH : LOW);
    if (bend_state)
        Serial.print("Water Level Threshold Exceeded, bend_value=");
    else
        Serial.print("Water Level Returned To Normal bend_value=");
    Serial.println(bend_value);
}
```

Note the first line of this code block: digitalWrite(ONBOARD_LED, bend_state ? HIGH : LOW);. This ternary operation polls the current state of the flex resistor based on the value (0 or 1) that we passed to the function. The conditional statement that follows prints out an appropriate message to the Arduino IDE's serial window. If the bend_state is true (HIGH), the flex resistor has been bent past the limits we defined. In other words, water has exceeded the threshold. If it's false (LOW), the flex resistor is straight (i.e., the water level is not rising).

All that is left to write is the program's main loop. Poll the FLEX_SENSOR pin (currently defined as analog pin 0) every second for any increase or decrease in value. When a flex event is detected, print the bend_value to the serial port so we can see it displayed in the Arduino IDE's serial window.

WaterLevelNotifier/WaterLevelSensor.pde
```
void loop()
{
    // wait a second each loop iteration
    delay(1000);
```

```
    // poll FLEX_SENSOR voltage
    bend_value = analogRead(FLEX_SENSOR);

        // print bend_value to the serial port for baseline measurement
        // comment this out once baseline, upper and lower threshold
        // limits have been defined
    Serial.print("bend_value=");
    Serial.println(bend_value);

    switch (bend_state)
    {
    case 0: // bend_value does not exceed high or low values
        if (bend_value >= FLEX_TOO_HI || bend_value <= FLEX_TOO_LOW)
        {
            bend_state = 1;
            SendWaterAlert(bend_value, bend_state);
        }
        break;
    case 1: // bend_value exceeds high or low values
        if (bend_value < FLEX_TOO_HI && bend_value > FLEX_TOO_LOW)
        {
            bend_state = 0;
            SendWaterAlert(bend_value, bend_state);
        }
        break;
    }
}
```

The main loop of the sketch will poll the value of the flex resistor every second. A switch statement tests the condition of the flex resistor. If its last status was straight (case 0:), check to see if it has since bent beyond the upper and lower threshold limits. If so, set the bend_state accordingly and call the SendWaterAlert function. Conversely, if the resistor's last status was bent (case 1:), check to see if it's now straight. If it is, set the bend_state variable to zero and pass that new state to the SendWaterAlert function.

Depending on the type of flex sensor and Ethernet shield used along with the voltage pin selected, your baseline value may be different from the baseline one I recorded. My flex sensor reported a value of 470.

Note the use of semicolons to mark the end of a line of instruction and brackets to identify conditional blocks. Save the file. It's also a good idea to place this and all other code you write under your preferred choice of version control before proceeding. I recommend Git,[6] but others like Mercurial and Subversion are certainly better than any non–version controlled alternative.

---

6. http://git-scm.com/

Later on, we will ask the SendWaterAlert function to call another function that will connect to a designated PHP server. This in turn will send an email alert that will contain the appropriate alert and the bend_value being monitored. But before we do, we will verify that our threshold test is working by monitoring the messages sent to the Arduino IDE's serial window.

Run the Sketch

Save and click the Verify button in the Arduino IDE's toolbar. This will compile the sketch to check for any syntax errors. After confirming that there are none, send the sketch to the Arduino by clicking the Upload button on the toolbar. You should see the Arduino's onboard LED flash a few times, indicating that it is receiving the sketch. When the rapid flashing stops, the sketch should be running.

Open up the Arduino IDE's Serial Monitor window. Assuming you haven't yet commented out the Serial.print("bend_value="); statement in the main loop of the sketch, observe the numbers that are continuously scrolling upward at a rate of roughly once a second on the serial monitor's display. If the characters being displayed in the window look like gibberish, make sure to select the correct baud rate (in this case, 9600) in the serial monitor's drop-down list located in the lower right corner of the serial monitor window. Take note of the values of the flex resistor when it is straight, bent forward, and backward.

Depending on the amount of electrical resistance and the type of hardware being used, update the FLEX_TOO_HIGH and FLEX_TOO_LOW constants to better calibrate them to the changing values you are seeing in the serial window. Once these defined amounts have been entered, save the program and upload again to the Arduino, performing the same procedure as before. It may take two or three tries to narrow in on the high and low values that help determine the bend state of the flex resistor.

With the modified upper and lower limits set to best suit your particular configuration, observe the Arduino's onboard LED to ensure that it lights up when the flex resistor bends far enough forward or backward and turns off when the resistor is straightened back to its original position.

Testing the Sketch

When you are confident that the hardware setup and the uploaded Arduino sketch are behaving correctly, it's time to try a simple water test by filling up a bowl with water and dipping the bobber into the water while holding the base of the flex resistor between your thumb and forefinger. As an extra precaution, wrap any exposed solder connecting the two wires to the flex resistor in waterproof electrical tape. I suggest wrapping the tape several layers thick, both to have a solid base to hold the resistor as well as to protect it from any errant drops of water that may accidentally splash or spill.

After properly and safely setting up the test, verify that as the buoyancy of the water deflects the bobber attached to the flex resistor, the resistor bends far enough in either direction to turn the LED light on.

Be careful not to submerge the exposed flex resistor. While the amount of current flowing through the Arduino is relatively low, water and electricity can make for a deadly combination. Place any electronics, including the flex resistor and attached bobber, in a sealed plastic bag with enough room to allow the flex resistor to bend. Use a high degree of caution to make absolutely sure to not get any of the exposed wiring or electrical connections wet. Doing so could damage your equipment or, even worse, you.

The base functionality of the water level notifier is complete. However, its method of communicating a rise in water height is limited to a tiny LED on the Arduino board. While that may be fine for science projects and people who work right next to the Arduino monitoring the water source in question, it needs to broadcast its alert beyond simple light illumination.

Receiving an email notification makes more sense, especially when the location of water measurement is somewhere in the home that is not frequently visited. Perhaps the detector will even operate at a remote location, such as when monitoring the status of a sump pit at a vacation home after a heavy rain.

To do so, we will need to clip on an Ethernet shield to the Arduino and write some code to send an email when the bend threshold is crossed. But before we add more hardware to this project, we first need to set up a web-based email notification application that our Arduino sketch can call upon when it needs to send out an alert.

3.5 Writing the Web Mailer

Libraries for sending email directly from the Arduino abound. But these all rely on a stand-alone, dedicated email server providing the mail gateway. So even though the mailer code can be compiled into the Arduino sketch, the solution still relies on an intermediary to send messages from the Arduino to the email inbox of the intended recipient(s).

If you have access to an SMTP mail server that you can connect to for outbound message transmission, check out Maik Schmidt's *Arduino: A Quick Start Guide* [Sch11]. His book supplies the necessary code and walkthrough on how to make this work. If you don't have access to a dedicated SMTP gateway, we can use an Internet web hosting service that supports sending email from a PHP script.

For this project, I have chosen a popular, preconfigured PHP-enabled web server with an SMTP outbound gateway, a configuration that popular website hosting companies like Dreamhost.net, Godaddy.com, and others offer to their customers.

The PHP script for sending email consists of only a few short lines of code. First, we will pass two parameters to the server: the type of alert to send and the recorded value of the flex resistor. Then we will compose a mail message containing the recipient's email address, the subject, and the message contents. Then we will send the email.

WaterLevelNotifier/wateralert.php
```php
<?php
// Grab the type of alert to email and
// the current value of the flex resistor.
$alertvalue = $_GET["alert"];
$flexvalue = $_GET["flex"];

$contact = 'your@emailaddress.com';

  if ($alertvalue == "1") {
  $subject = "Water Level Alert";
  $message = "The water level has deflected the flex
             resistor to a value of " . $flexvalue . ".";
  mail($contact, $subject, $message);
  echo("<p>Water Level Alert email sent.</p>");
  } elseif ($alertvalue == "0") {
  $subject = "Water Level OK";
  $message = "The water level is within acceptable levels.
             Flex resistor value is " . $flexvalue . ".";
  mail($contact, $subject, $message);
  echo("<p>Water Level OK email sent.</p>");
  }

?>
```

The script calls the built-in PHP mail function that passes three required parameters: recipient(s), subject, and the body of the email. Yes, it's that simple.

Save the code to a file called wateralert.php in the root web directory of your PHP server. You can test the script by opening your web browser and visiting *http://MYPHPSERVERNAME/wateralert.php?alert=1&flex=486*. The page should return a Water Level Alert email sent. message in the browser window, and

> **Why Use a PHP-Enabled Web Server for This Project?**
>
> Quite simply, because they are the most prevalent web-hosting server configurations. While I personally prefer a more modern web application framework like Django or Ruby on Rails hosted within a virtual private server (VPS), these technologies are not as universally supported by hosting providers compared to PHP. This wouldn't be a problem if we hosted the web server within our own network (which we in fact do in Chapter 7, *Web-Enabled Light Switch*, on page 105) or had access to a VPS. But given the setup configuration overhead associated with running both a web server and an email server that correctly sends outbound SMTP messages, it's easier to go this route for our first project.
>
> Sending email via PHP can be done with a single PHP file in a single line of code. That said, if you are interested in writing functional equivalents for your personal favorite web frameworks, go for it! If you succeed, please considering sharing your discoveries with the Programming Your Home book discussion community.

a corresponding email message should appear in the defined recipient's inbox. If it doesn't, check your PHP server settings and make sure that your web server is properly configured to use a working email gateway. If you're still not having luck with the message test, contact your website hosting provider to make sure your hosted solution is correctly configured for PHP email messaging.

By abstracting the delivery mechanism from the logic running in the Arduino, we can easily modify the message recipients and contents.

Now that we have a working message gateway, we can hook up the Arduino to an Ethernet shield so the deflected flex resistor can talk to the rest of the world.

3.6 Adding an Ethernet Shield

Attach the Ethernet shield to the Arduino by lining up the base pins so that the Ethernet jack is on top and facing the same direction as the Arduino USB jack. Reconnect the wires to the 5V and analog-in 0 (A0) pins found on the Ethernet shield just like you did when these wires were connected to the Arduino.

Do the same for the 10k ohm resistor bridging across the ground (Gnd) and A0 pins. Run your test again and check the values. In my tests, the base value being read was different compared to the Arduino without the Ethernet shield, and yours will likely reflect similar results. Since we're more interested in the deviation from this base value than the calibration of the actual value

> **Securing Your Notifications**
>
> If you plan on having this PHP script provide a permanent service for Arduino message passing, consider adding a layer of security to the transmission signal so that only the Arduino can trigger the message condition.
>
> This could be done by something as simple (though weak) as a password value passed in the HTTP GET parameters or by a more secure hash transaction that trades an authentication conversation between the Arduino and the web server. While adding a good security routine is beyond the scope of this project, it's a good idea to incorporate such functionality so that your publicly exposed PHP email entry point isn't abused by unwelcome connections.

itself, it's important to use the unbent resistor value in the code and then determine how far of a plus or minus deflection from this base value is acceptable before transmitting the alert.

Now that our hardware is network-enabled, we can add the necessary code to our sketch that transmits the flex sensor status to our PHP server.

Coding the Shield

We will programmatically send data via the Ethernet shield. But we first must include a reference in the sketch to both the Arduino Ethernet library and its dependency, the Serial Peripheral Interface (SPI) library.[7] These two libraries contain the code needed to initialize the Ethernet shield and allow us to initialize it with network configuration details. Both libraries are included in the Arduino IDE installation, so the only thing we need to do is import the SPI.h and Ethernet.h libraries via the #include statement. Add these statements at the beginning of the sketch:

WaterLevelNotifier/WaterLevelNotifier.pde
```
#include <SPI.h>
#include <Ethernet.h>
```

With the Ethernet library dependency satisfied, we can assign a unique Media Access Control (MAC) and IP address to the shield. While DHCP libraries are available from the Arduino community, it's easier just to set the shield with a static IP address.

For example, if your home network uses a 192.168.1.1 gateway address, set the address of the shield to a high IP address like 192.168.1.230. If you plan

---

7. http://arduino.cc/en/Reference/Ethernet and http://www.arduino.cc/playground/Code/Spi, respectively.

> ### Arduino on Linux and the Ethernet Library
>
> If you are using the Linux version of the Arduino IDE, you might encounter a problem with the Ethernet reference library. The problem manifests itself by transmitting garbled broadcasts from the Ethernet shield. Fortunately, a fork of the Ethernet library, aptly named Ethernet2, is available for download.[a] Refer to Appendix 1, *Installing Arduino Libraries*, on page 209, for more details. Once the Ethernet2 library is installed, replace the broken Ethernet.h in the original #include statement with #include Ethernet2.h instead.
>
> a. http://code.google.com/p/tinkerit/source/browse/trunk/Ethernet2+library/Ethernet2/

on using this address as a persistent static IP, refer to your home router's documentation on how to set a static IP range within a DHCP-served network.

WaterLevelNotifier/WaterLevelNotifier.pde
```
// configure the Ethernet Shield parameters
byte MAC[] = { 0xDE, 0xAD, 0xBE, 0xEF, 0xFE, 0xEF };

// replace this shield IP address with one that resides within
// your own network range
byte IPADDR[]  = { 192, 168, 1, 230 };

// replace with your gateway/router address
byte GATEWAY[] = { 192, 168, 1, 1 };

// replace with your subnet address
byte SUBNET[]  = { 255, 255, 255, 0 };

// replace this server IP address with that of your PHP server
byte PHPSVR[] = {???, ???, ???, ???};

// initialize a Client object and assign it to your PHP server's
// IP address connecting over the standard HTTP port 80
Client client(PHPSVR, 80);
```

Assign constants for the static MAC and IP addresses that will be used by the Ethernet shield. Add the address of your Internet router to the GATEWAY value, and add your SUBNET value as well (most home network subnets are 255.255.255.0). The IP address of your PHP server also has to be declared prior to the sketch's setup routine.

With the constants declared, we can now properly initialize the Ethernet shield in the setup section of the sketch.

WaterLevelNotifier/WaterLevelNotifier.pde
```
void setup()
{
    // for serial window debugging
```

> ### Ethernet Shield DNS and DHCP
>
> The Ethernet library does not natively include any DNS or DHCP functionality. This capability is expected to arrive in an upcoming release of the Arduino platform. But until that day arrives, we cannot use a server name like www.mycoolwaterlevelproject.com for a web server address and must use the server's assigned IP address instead.
>
> Thanks to the efforts of Arduino enthusiast George Kaindl, using DNS and DHCP with an Ethernet shield is possible. If you don't mind the extra overhead these libraries add to the Arduino's already constrained program storage capacity, check his Arduino Ethernet libraries for more details.[a]
>
> ---
>
> a. http://gkaindl.com/software/arduino-ethernet

```
  Serial.begin(9600);

  // set up on board led on digital pin 13 for output
  pinMode(ONBOARD_LED, OUTPUT);

  // Initialize Ethernet Shield with defined MAC and IP address
  Ethernet.begin(MAC, IPADDR, GATEWAY, SUBNET);
  // Wait for Ethernet shield to initialize
  delay(1000);
}
```

Note the use of the Ethernet object in Ethernet.begin(MAC, IPADDR, GATEWAY, SUBNET);. This is where the Ethernet shield gets initialized with the assigned Media Access Control (MAC) address and IP Address.

OK, we have a working network connection. Now we can move on to the next step of requesting the appropriate emailer page on your PHP server when the bend thresholds have been exceeded.

Sending a Message

Up to this point, we have told the Arduino to report the analog values being generated by the flex resistor, initialized the Ethernet shield to connect the Arduino to our network, and added stubs for routines to call out to our PHP server script. Now it's time to add that routine. We'll call it ContactWebServer.

The ContactWebServer routine will take the same two parameters we captured for the SendWaterAlert function, namely band_value and bend_state. Add the ContactWebServer(bend_value, bend_state); line at the end of the SendWaterAlert function, since we will talk to the designated PHP web server address each time the flex resistor state changes.

We're almost done. We just have to write the body of the `ContactWebServer` function. This will consist of connecting to the PHP web server and printing the well-formed HTTP GET string to the server. The string will contain and pass the values of the `bend_state` and `bend_value` variables. These will then be parsed on the server side and the PHP function will respond in kind.

```
WaterLevelNotifier/WaterLevelNotifier.pde
void ContactWebServer(int bend_value, int bend_state)
{
    Serial.println("Connecting to the web server to send alert...");

    if (client.connect())
    {
        Serial.println("Connected to PHP server");
        // Make an HTTP request:
        client.print("GET /wateralert.php?alert=");
        client.print(bend_state);
        client.print("&flex=");
        client.print(bend_value);
        client.println(" HTTP/1.0");
        client.println();
        client.stop();
    }
    else
    {
        Serial.println("Failed to connect to the web server");
    }
}
```

It's time to test the completed sketch. Download it to the Arduino, open up a serial monitor window, bend the flex resistor, and watch the messages. Check your recipient's inbox for the corresponding email messages. Did you receive the "Water Level Alert" and "Water Level OK" email messages that correspond to the notifications you saw in the serial monitor window? If not, make sure that your Arduino is connected to your home network by pinging the IP address you assigned.

Test the PHP email URL and verify that you receive an email when you enter *http://MYPHPSERVER/wateralert.php?alert=1&flex=486* into your web browser. When everything works as expected, we will be ready to put the finishing touches on this project and make it fully operational.

3.7 All Together Now

We're nearing the home stretch. Your hardware should look like the setup pictured in Figure 4, *An assembled water level notifier*, on page 41. All that remains is mounting the flex resistor securely and safely in place so that its flexion is accurately detected and not impeded by any obstacles.

The base where the two wires attach to the exposed flex resistor leads needs to be firmly stabilized so that when the water level rises and pushes the bobber upward, the base does not pivot at its fulcrum. If it does pivot, the flex resistor will remain straight and the running Arduino sketch will fail to send the appropriate alert notification. Keep the base stabilized and prevent it from pivoting.

Try using hot glue, heat shrink tubing, or duct tape. If the base still moves, try attaching a small wood chip splint on each side of the base of the flex resistor. Extend the splint length-wise approximately two centimeters above and below the base. Then snugly wrap the splint several times with electrical tape. Tack the top of the splinted base to a small wood post (such as that cut from a typical two-by-four piece of lumber) that spans the diameter of the hole containing the water source.

In the case of a sump pit, you will need to remove the cover of the pit, measure the interior diameter and visit a lumberyard or hardware store that can cut the wood for you. Add an extra centimeter to the cut so that the beam can be wedged tightly as it spans the pit.

Similar principles apply in the case of a dehumidifier. Instead of using a large piece of wood to act as the mounting base support, use the bottom, pants-hanging portion of an old wooden hanger that can be cut to slightly longer than the diameter of the dehumidifier's water collection bucket. Mount the base of the splinted flex resistor in the center of the wood support. Depending on the depth of the dehumidifier's bucket, you may need to raise the base of the flex resistor higher so that the alert doesn't trigger prematurely when the bucket is only half-full.

Once you're satisfied with the stability of the mounted resistor, place the bobber and flex resistor inside a small plastic bug, such as a locking seal sandwich bag. This will keep the resistor dry and protected if the water level rises excessively. Run the wires attached to the resistor a meter or more from the measured water source and attach them to the Arduino/Ethernet shield assembly. Power the Arduino using the 9-volt power supply and attach the network cable to the Ethernet shield. Several seconds after you power up the Arduino, perform a quick bend test. If you received the water alert and all-clear messages in your email inbox, then you have succeeded!

Replace the cover of the water containment vessel you are monitoring and wait for your device to alert you to rising water levels.

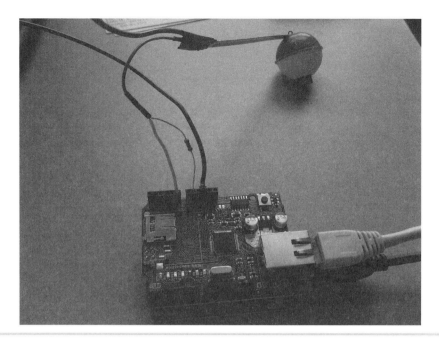

Figure 4—An assembled water level notifier

3.8 Next Steps

Congratulations on completing the first Arduino-assisted home automation project in this book. You have already learned a lot of reusable ideas in this project. You programmed an Arduino, captured and processed data from a flex resistor, and sent emails with the help of a PHP-enabled web server and the Arduino Ethernet shield. We will be applying these concepts again in some of the other projects in this book.

The cool thing about designing and building your own projects is that they can each be tailored to your own exacting requirements. Prefer a tweet instead of an email alert when the water level exceeds the measurement threshold? No problem. Replace the email functionality with the code from the Tweeting Bird Feeder project later in this book. Want an overt visual indicator instead of an electronic message, something like a blockbuster action movie warning lamp that flashes on and off? Easy. Hook up a switch to the lamp that can be controlled to turn on and off at regular intervals with code lifted from the web-enabled light switch project.

Here are a few other ideas to further extend the use of a flex resistor in the home:

- Use the variable analog data that is emitted from the resistor to determine not only when it has been flexed but also to what degree. This could be useful in a rain gauge application used to track incremental measurements of rainfall based on the deflection of the resistor by the buoyant bobber.

- Add an hourly data transmission to the sketch and a routine in the PHP component to receive the message. Current bend values should be transmitted in this message as well. Check the values for anomalies, such as having no value (0) if it's broken or something greater than 999 if there is a short circuit. Send an email alert when such threshold values are detected. Additionally, if the transmission isn't received in a two-hour time frame, send an email informing the recipient of that fact. This enhanced monitoring will let you know that your hardware may be having issues and needs further attention.

- Temperature variations may affect the calibration of the flex sensor. Attach a temperature sensor and dynamically change the trigger point values based on the surrounding ambient temperature readings.

- Concerned about losing roofing tiles, shingles, or siding to the wind? Replace the bobber with a wind cup like those found mounted on weather stations sold by scientific instrument supply companies, set it up outside, and receive an email alert when the wind is becoming excessively strong.

- If you use a flap door for your pet, anchor one end of the flex resistor to the flap frame and slide the untethered end into a small vinyl tube attached to the flap to allow the resistor to slide freely but still flex when the door flap is being pushed open on either end. Combine the sensor trigger with a web cam capture so you can verify that it's your family pet coming in and out of the house and not some uninvited guest.

CHAPTER 4

Electric Guard Dog

Remember the last time you visited a home with a big dog? Did hearing the canine barking at the sound of a doorbell make you think twice before entering the premises? Most dog owners appreciate the vigilant home surveillance that their pets provide. These furry friends have a knack of detecting motion and springing immediately into action, barking and bumping their snouts against window curtains and doors in hyperactive effort to see who or what is outside.

With the Electric Guard Dog, you will be able to derive a similar security benefit minus the hassles of cleaning up dog hair afterward (Figure 5, *Deter unwanted visitors with the Electric Guard Dog*, on page 44).

This project combines the Arduino board with a wave shield, a Passive InfraRed (PIR) sensor, and a servo motor. When programmed and activated, the assembly will give the illusion of an angry dog eager to pounce on an unwanted trespasser. A small rod attached to the arms of a servo motor will bob up and down when the servo rotates. A wad of cotton cloth attached to the other end of the rod will be positioned against a window curtain. When motion is detected, the servo will rotate, moving the rod up and down. The cloth attached to the other end of the rod will bump against the curtain in time with random barks and growls coming from a speaker plugged into the wave shield. This sound and motion will give the illusion of a noisy dog trying to poke and prod with its nose behind a door or window curtain.

The completed project is fully portable, since the Electric Guard Dog can be positioned in any doorway, window, or room that you want to get someone or something's attention when the motion detector is triggered.

Figure 5—Deter unwanted visitors with the Electric Guard Dog.

4.1 What You Need

This project requires only a few components. The total cost for all the parts should be under a hundred dollars. But since all the parts can be reused in other projects in this book and in future DIY efforts, it is a very reasonable investment consideration. To construct an Electric Guard Dog, you will need the following (see Figure 6, *Electric Guard Dog parts*, on page 45):

1. An Arduino Uno
2. An Adafruit music and sound add-on pack for Arduino (includes wave shield, speaker, wire, and SD card)[1]
3. A high-torque standard servo[2]
4. A Passive InfraRed (PIR) motion sensor
5. A 9-volt power supply to power the Arduino once untethered from the USB development cable
6. A sturdy wooden rod with cotton or rubber affixed to the tip to serve as a surface-protecting end-cap
7. Wire, twist ties, or rubber bands to affix the wooden rod to the servo gear

---

1. http://www.adafruit.com/products/175
2. http://www.adafruit.com/products/155

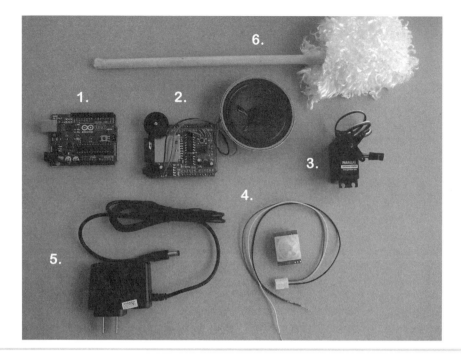

Figure 6—Electric Guard Dog parts

You will also need a standard A-B USB cable to connect the Arduino to the computer. The servo can be purchased at a local hobby shop, and the PIR can be purchased from a number of electronic parts retailers, including Fry's and Radio Shack, as well as from online electronics retailers like Adafruit or Sparkfun.

Let's start by connecting the project's three main components to make them collectively act in a more aggressive manner.

4.2 Building the Solution

This is one of the easier projects in the book, since it relies entirely on the Arduino, an add-on shield, a sensor, and a servo motor. When constructed, the completed assembly should look similar to the one shown in Figure 7, *An Electronic Guard Dog*, on page 47. Here's how we will build it:

1. Attach an Adafruit wave shield to the Arduino.

2. Connect a PIR to the wave shield's power, ground, and one of the available digital pins.

> **Joe asks:**
> ### Is There an Arduino Shield That Can Play MP3 Files?
>
> Yes! Electronics project retailer Sparkfun sells an Arduino board called the MP3 shield that is similar in function to Adafruit's wave shield.[a] However, due to the differences in the libraries used, I will focus on the wave shield implementation and leave it to our more adventurous readers to pursue Sparkfun's MP3-based alternative on their own. And for those who need an audio shield that plays even more sound file formats like Windows Media Audio, MIDI, and Ogg Vorbis, the Maker Shed sells the Seeed Music Shield, which nicely integrates audio file playback capabilities in a well-designed shield.[b]
>
> ---
>
> a. http://www.sparkfun.com/products/9736
> b. http://www.makershed.com/ProductDetails.asp?ProductCode=MKSEEED14

3. Connect a servo to the wave shield's power, ground, and another one of the available digital pins.

4. Download additional Arduino libraries that allow the wave shield to be easily controlled while preventing resource conflicts with sending instructions simultaneously to the servo.

5. Write a sketch that randomly moves the servo and plays back a snippet of audio when motion is detected by the PIR.

If you haven't already assembled and tested your wave shield, follow Ladyada's instructions on how to do so.[5] When you have confirmed that it works, we can enhance the board by attaching the PIR sensor and servo motor actuator to the available wave shield's pins.

4.3 Dog Assembly

Take a look at the schematic in Figure 8, *Wiring diagram for the Electric Guard Dog*, on page 48. The graphic shows wiring plugging into the wave shield. The wave shield is stacked on top of the Arduino board. Note that the wave shield uses several of the pins for its own use to interact with the Arduino, which is why not all passthrough pins are available for the sketch. Closely follow the wiring diagram and you should not have a problem.

---

5. http://www.ladyada.net/make/waveshield/

Figure 7—An Electronic Guard Dog

Attach the positive lead of the PIR to the 3.3v pin on the wave shield. Connect the negative lead to one of the wave shield's available ground pins. Then attach the control wire (the middle pin/wire on the PIR) to the wave shield's digital pin 12.

Next, attach the servo's positive wire to the wave shield's 5v pin. Connect the negative lead to the wave shield's other available ground pin. Finally, connect the control wire to the wave shield's digital pin 11.

For brief testing purposes, you can attach male pins to the wires and plug them directly into the sockets on the wave shield. More reliable connections can be achieved by using either male or female header pins instead. These can be obtained directly from various Arduino board suppliers. If you plan on using the wave shield exclusively for this project, you can solder the wiring permanently to the shield for the most stable electrical connection possible.

There is one more step we should take before writing the sketch. We need to either record and digitize a dog growling and barking in various ways or legally download audio samples from the Internet of snarling, barking dog sounds.

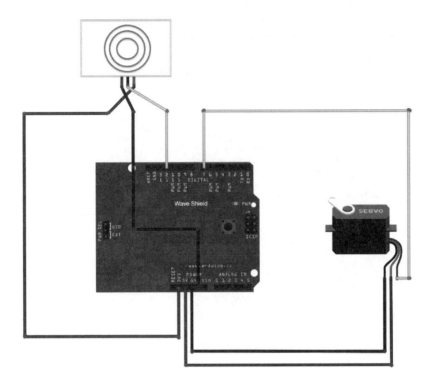

Figure 8—Wiring diagram for the Electric Guard Dog

The first option takes more time and requires access to a big dog that can bark, snarl, and growl on command—with a microphone near its toothy yapper, no less! While this requires a bit more extra work, the results produce a more consistent and realistic effect. And because you know the source, playback generates a more meaningful audio cue.

The second option of searching on the Internet for a variety of angry dog audio samples is more convenient but rarely produces a consistent and believable overall effect. This is especially true when the samples are acquired from a variety of dog breeds. How can a dog have the toothy snarl of a Doberman one minute and the yapping of a miniature poodle the next? Also, downloading audio samples from the Internet has copyright implications that have to be respected. One website that I recommend visiting is the Freesound Project,[7]

---

7. http://www.freesound.org

> **Joe asks:**
> ### How Does a PIR Sensor Work?
>
> A PIR detects motion by comparing two samples of infrared radiation being emitted by a body warmer than the background environment it is moving against. When either side of the sensor detects a greater value than the other, it sends a signal to the digital out pin that motion has been detected. The IR sensor at the heart of a PIR is typically covered by a dome-shaped lens that helps to condense and focus light so that it is much easier for the sensor to detect infrared variations, and thus, motion.
>
> For a more detailed explanation of the theory behind PIRs, visit Ladyada's informative web page on the subject.[a]
>
> ---
> a. http://www.ladyada.net/learn/sensors/pir.html

which features a number of samples available under the Creative Commons Sampling Plus license.

After you have obtained five audio clips using either approach, you need to convert them to a format the wave shield can interpret. Based on the conversion instructions on Ladyada's website,[8] samples must not exceed a 22KHz 16-bit mono PCM (WAV) format. You want the highest audio quality possible, and there should be plenty of space on the SD card to store them. The audio clips you select for the project should not exceed five seconds in duration so they appear more synchronized with the servo motion when the audio is played back.

You can use an audio editor like Audacity to import and convert and save your audio clips to the correct format.[9] Make sure they are compatible by copying the converted files to the wave shield's SD card and running the dap_hc.pde sketch posted on Ladyada's website.[10] Note that we're going to make one change to Ladyada's wave shield demo sketch. Instead of the newer wavehc library it uses, we are going to use the older AF_Wave library. That way, we can use Arduino community forum member avandalen's MediaPlayer library[11]—it makes working with wave shield sound files far easier. We will take a closer look at this library and another Arduino community contributor's library for servos when we write the sketch in the next section.

---

8. http://www.ladyada.net/make/waveshield/convert.html
9. http://audacity.sourceforge.net/
10. http://www.ladyada.net/make/waveshield/libraryhc.html
11. http://www.arduino.cc/playground/Main/Mediaplayer

4.4 Dog Training

The sketch we write will monitor the PIR for any motion events. If movement is detected, the shield will randomly play one of five different audio files stored on the wave shield's SD card. Simultaneously, the servo motor rotates up to 150 degrees, depending on the sound effect being played back. Attach a wooden rod to the servo gear and the servo's rotation will move the rod up and down. When the rod is positioned behind a curtain, it will give the illusion of a dog's snout attempting to nudge the curtain aside so it can see who's at the door or window.

To begin, we need to include the MediaPlayer.h header file along with its two dependencies, pgmspace.h (part of a memory management library included in the Arduino's standard installation) and util.h (part of the original wave shield's AF_Wave library). Because the MediaPlayer class relies on the AF_Wave library, make sure you have already downloaded, unzipped, and copied the uncompressed AF_Wave folder into the Arduino's libraries folder.[12]

Next, create a new sketch in the Arduino IDE called ElectricGuardDog. Download the MediaPlayer library from the Arduino playground website;[13] extract the zip archive; and place the unzipped MediaPlayer.h, MediaPlayer.pde, and MediaPlayerTestFunctions.pde files into the ElectricGuardDog folder created by the Arduino IDE when it created the ElectricGuardDog.pde file. If you downloaded the project files for the book, the Mediaplayer library file dependencies have already been pre-bundled for you. The Mediaplayer library allows us to control audio file playback very easily.

We will also need to call upon another custom library to operate the servo motor. If you try to compile the sketch using the standard Arduino Servo class, the program will fail with this error:

```
Servo/Servo.cpp.o: In function `__vector_11':
/Applications/Arduino.app/Contents/Resources/Java/libraries/Servo/Servo.cpp:103:
multiple definition of `__vector_11'

AF_Wave/wave.cpp.o:/Applications/Arduino.app/
Contents/Resources/Java/libraries/AF_Wave/wave.cpp:33: first defined here
```

What's going on here? The AF_Wave library is taking over the vector interrupt as the standard Servo library. Fortunately for us, Arduino community contributor Michael Margolis has written a library that gives the Arduino the ability

---

12. http://www.ladyada.net/media/wavshield/AFWave_18-02-09.zip
13. http://www.arduino.cc/playground/Main/Mediaplayer

to control up to eight servo motors simultaneously. By doing so, his library also circumvents the duplicate resource problem exhibited by the original Servo library when combined with a wave shield.

Download the ServoTimer2 library,[14] unzip it, and copy the ServoTimer2 folder into the Arduino libraries folder. Keep in mind that each time you add a new library to the Arduino libraries folder, you need to restart the Arduino IDE so the Arduino's avr-gcc compiler will recognize it.

After the wave shield's AF_Wave and servo motor's ServoTimer2 library dependencies have been satisfied, add these references to the beginning of the sketch:

ElectricGuardDog/ElectricGuardDog.pde
```
#include <avr/pgmspace.h>
#include "util.h"
#include "MediaPlayer.h"
#include <ServoTimer2.h>
```

Create several variables to store Arduino pin assignments and sensor/actuator starting values.

ElectricGuardDog/ElectricGuardDog.pde
```
int ledPin        = 13;  // on board LED
int inputPin      = 12;  // input pin for the PIR sensor
int pirStatus     = LOW; // set to LOW (no motion detected)
int pirValue      = 0;   // variable for reading inputPin status
int servoposition = 0;   // starting position of the servo
```

Next, create two objects constructed from the MediaPlayer and ServoTimer2 libraries to more easily manipulate the servo motor and audio playback.

ElectricGuardDog/ElectricGuardDog.pde
```
ServoTimer2 theservo;      // create servo object from the ServoTimer2 class
MediaPlayer mediaPlayer;   // create mediaplayer object
                           // from the MediaPlayer class
```

Assign the variables we created to the Arduino pinModes in the sketch's setup() routine. Establish a connection to the Arduino IDE serial window to help monitor the motion detection and audio playback events. Call the Arduino's randomSeed() function to seed the Arduino's random number generator. By polling the value of the Arduino's analog pin 0, we can generate a better pseudorandom number based on the electrical noise on that pin.

ElectricGuardDog/ElectricGuardDog.pde
```
void setup() {
  pinMode(ledPin, OUTPUT);    // set pinMode of the onboard LED to OUTPUT
  pinMode(inputPin, INPUT);   // set PIR inputPin and listen to it as INPUT
```

---

14. http://www.arduino.cc/playground/uploads/Main/ServoTimer2.zip

```
  theservo.attach(7);      // attach servo motor digital output to pin 7
  randomSeed(analogRead(0)); // seed the Arduino random number generator
   Serial.begin(9600);
}
```

With the library, variable, object, and setup initialization out of the way, we can now write the main loop of the sketch. Essentially, we need to poll the PIR every second for any state changes. If the PIR detects motion, it will send a HIGH signal on pin 12. When this condition is met, we power the onboard LED and send a motion detection message to the Arduino IDE's serial window.

Next, we generate a random number between 1 and 5 based on the seed we created earlier. Based on the value generated, we then play back the designated audio event and move the servo motor a predefined amount of rotation. After that, we wait a second before returning the servo to its starting position and run the loop again. If the PIR fails to detect motion (that is, if the signal on pin 12 is LOW), we turn off the onboard LED, send a No motion message to the serial window, stop the audio playback, and set the pirStatus flag to LOW.

ElectricGuardDog/ElectricGuardDog.pde
```
void loop(){
  pirValue = digitalRead(inputPin); // poll the value of the PIR
  if (pirValue == HIGH) {          // If motion is detected
    digitalWrite(ledPin, HIGH);         // turn the onboard LED on
    if (pirStatus == LOW) {                  // Trigger motion
      Serial.println("Motion detected");

      // Generate a random number between 1 and 5 to match file names
          // and play back the file and move the servo varying degrees
      switch (random(1,6)) {
        case 1:
          Serial.println("Playing back 1.WAV");
          theservo.write(1250);
          mediaPlayer.play("1.WAV");
          break;
        case 2:
          Serial.println("Playing back 2.WAV");
          theservo.write(1400);
          mediaPlayer.play("2.WAV");
          break;
        case 3:
          Serial.println("Playing back 3.WAV");
          theservo.write(1600);
          mediaPlayer.play("3.WAV");
          break;
        case 4:
          Serial.println("Playing back 4.WAV");
          theservo.write(1850);
          mediaPlayer.play("4.WAV");
```

```
        break;
      case 5:
        Serial.println("Playing back 5.WAV");
        theservo.write(2100);
        mediaPlayer.play("5.WAV");
        break;
    }

    delay(1000);        // wait a second
    theservo.write(1000); // return the servo to the start position
    pirStatus = HIGH;   // set the pirStatus flag to HIGH to stop
                        // repeating motion
  }
} else {
  digitalWrite(ledPin, LOW); // turn the onboard LED off
  if (pirStatus == HIGH){
    Serial.println("No motion");
    mediaPlayer.stop();
    pirStatus = LOW;     // set the pirStatus flag to LOW to
                         // prepare it for a motion event
  }
 }
}
```

Save the code as ElectricGuardDog.pde and open up the newly created ElectricGuardDog folder containing the ElectricGuardDog.pde source file. Place the unzipped MediaPlayer files into the ElectricGuardDog directory. Double-check that the uncompressed ServoTimer2 library files are in the Arduino libraries directory.

Reopen the Arduino IDE, load up the ElectricGuardDog.pde file, and click the Verify icon in the Arduino IDE toolbar. If everything compiled without errors, you have entered the code correctly and placed the dependent library files in the correct locations. If not, review the error messages to see what dependencies may be missing and correct accordingly.

With the sketch compiled successfully, we're ready to test and tweak the code.

4.5 Testing It Out

Place the PIR sensor at a convenient location to test motion detection, download the sketch, and open the Arduino IDE's serial window.

Trigger the PIR sensor by waving your hand in front of it. Your guard dog should react with a random audio clip and servo motion. If you want the servo motor to rotate differently, modify theservo.write() method calls with values ranging from 1000 to 2200. This is because the ServoTimer2 library uses microseconds instead of the angle of degrees used by the original Servo library to measure pulse widths. As a result, you may need to experiment to find the right degree of movement. After getting the hang of the timing based on the size of the servo you are using, determining the ideal values to elicit the desired amount of rotation will become second nature.

Now that you have tested and tweaked the servo timing synchronized with the appropriate audio clip, it's time to put the finishing touches on the final placement of the hardware.

4.6 Unleashing the Dog

Consider where the PIR should be mounted. Placing it behind a glass windowpane to track outdoor movement will not work since the detector cannot analyze infrared signatures. Ideally, the PIR should be placed in the unobstructed line of sight of the area being monitored. If it's just outside your front door, thread wiring from the PIR mounted above the door to the Arduino/wave shield mounted in an enclosure inside the house.

Play with the playback audio level. The small speaker that accompanies Adafruit's music and sound add-on pack may be adequate for testing, but it's hardly loud enough to get the attention of anyone in the next room (let alone someone who is outdoors). Use the wave shield's headphone jack and connect it to a powered speaker, such as a boom box or home stereo. Set the volume loud enough to get a visitor's attention.

Attach one end of the wooden rod to the servo wheel using wire, twist ties, or rubber bands. Cover the other end of the rod with cloth, a cotton ball, or a rubber cap to protect the surface that the rod's tip will be bumping against. You can further embellish the wire frame with a plastic dog snout from a costume store. Get creative! Just be sure not to attach something so heavy that the servo cannot generate enough torque to adequately move the attached wire.

Place the servo assembly next to a window curtain, preferably near the entryway. When the PIR is triggered and the barking audio is played back, the faux appendage will nudge the curtain and give the illusion of a dog's nose moving behind the curtain. From the visitor's point of view, it will look like an agitated animal is just behind the door, waiting to pounce. It will take some tweaking to get right, but once your setup is properly configured, the motion-detected playback events should look and sound very convincing!

4.7 Next Steps

Here are a couple of other ideas to elevate this project to the next level:

- Replace the dog barking samples with a booming klaxon, a piercing alarm bell, or science fiction self-destruct sound effects. Swap out the fake dog snout attached to the servo arm with a laser pen light that sweeps the entryway. Imagine your front doorway looking like something out of a science fiction thriller!

- Add an ultrasonic rangefinder and alter the reaction of sound and motion based on the proximity of the movement being captured. As a intruder comes closer to the sensor, have the volume get louder and the servo move more erratically. The closer one gets to the sensor, the more agitated the Electric Guard Dog becomes.

- Upscale your Guard Dog into a super-sized, weatherized garden scarecrow. Use more powerful stepper motors connected to a higher voltage external power source. Make a life-size replica of yourself in old coveralls and use PVC tubing connected to the stepper motors and wire akin to the strings of a marionette to control the excited motions of the scarecrow's arms and legs.

- Reuse your Electric Guard Dog rig on Halloween. Cover in a "ghost" sheet or configure a mask with hinged jaws attached to servos that greets visitors with a spooky voice and ethereal movements.

- Combine the Electric Guard Dog with other projects from the book to turn on lights, send an email, or lock/unlock the door when motion is detected.

CHAPTER 5

Tweeting Bird Feeder

Both of my kids are bird lovers. They have had parakeets since they were toddlers and enjoy watching wild birds nest and feed outside their bedroom windows. But one of the chores that somehow always slips past us is refilling the feeders with birdseed. For a variety of reasons, there may be days, sometimes even weeks, that go by without a refill. Wouldn't it be so much easier for the feeder to tell us when it needed to be refilled?

That need was the genesis of inspiration for this project, and what better way to receive the notification than via a tweet on Twitter. Interested friends and family members can follow the feeder and know when birds are feeding from it, when it needs a refill, and when the refill chore has been satisfied. (See Figure 9, *Receive a Twitter notification from your bird feeder*, on page 58.)

Since we will already be tracking the refilling patterns via Twitter, let's make the feeder broadcasts even more interesting by adding a homemade sensor on the feeder perch that will record when birds are enjoying a meal and for how long. Before posting the tweets, we will timestamp and record these events to a database so we can visualize feeding patterns over time.

Was April a more ravenous month for the birds compared to July? Are early mornings a busier time of day than late afternoons? What is the average time birds perch at the feeder? What are the time intervals between perches? How frequently does the feeder need to be refilled with seed? With the Tweeting Bird Feeder, you will be able to take on the role of field researcher to discover these and other feeding behavior questions. It's time to fly!

5.1 What You Need

Since this will be our first outdoor project, the equipment costs are more expensive for several reasons. First, unless you are willing to drill holes

Figure 9—Receive a Twitter notification from your bird feeder ...when birds are perching, as well as when seed needs replenishing.

through your walls or leave a window or door open to run an Ethernet cable to the feeder, we will need an untethered way to broadcast sensor events. Fortunately, a low-power and relatively low-cost option exists in the form of XBee radios. While these take a little extra effort to configure initially, they are fairly reliable, easy to communicate with, and don't require much attention once operational.

Second, while we could use the standard size Arduino Uno (as shown in the wiring diagrams throughout this chapter), it might prove to be too long and/or too wide to conveniently fit into a typical bird feeder. Consequently, I recommend spending a few extra dollars on an Arduino Nano. The Nano is better suited to match the feeder's space constraints. The nice thing about the Nano is that the pin configurations and the hardware are nearly identical to that of its bigger brother, and the Nano delivers all that Arduino goodness in a package within a much smaller footprint.

Third, while it is feasible to power the electronics via a long extension cord plugged into an outdoor outlet akin to a holiday lighting scenario, such a

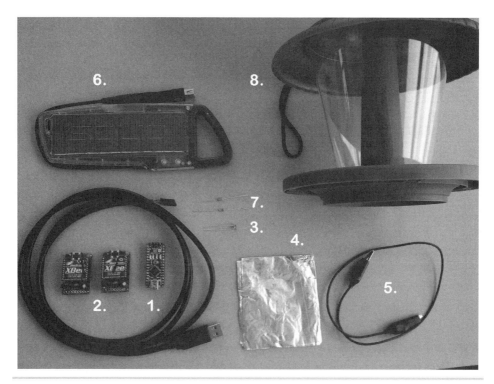

Figure 10—Tweeting Bird Feeder parts

configuration is not a self-contained system. Besides, it would be appropriate to incorporate a greener energy option to be kinder to our environment.

Finally, due to the need to protect the electronics from the elements, we will need to do a good job weatherizing our assembly. Here's the shopping list (refer to the photo in Figure 10, *Tweeting Bird Feeder parts*, on page 59):

1. An Arduino Uno or an Arduino Nano[1]
2. Two XBee radios with adapter kits and FTDI connector cable[2]
3. A photocell
4. A strip of aluminum foil
5. A piece of wire
6. A small solar panel with built-in rechargeable battery and USB connector, such as those provided by Solio[3]

---

1. http://www.makershed.com/ProductDetails.asp?ProductCode=MKGR1
2. http://www.adafruit.com
3. http://www.solio.com/chargers/

7. One 10k ohm resistor and one 10M ohm resistor—verify that the color bands on the resistors are brown, black, orange, and gold for the 10k ohm and brown, black, blue, and gold for the 10M ohm resistors. Refer to Figure 11, *Tweeting Bird Feeder resistors*, on page 61. Also shown in the photo is the photocell (also referred to as a CdS photoresistor).
8. A bird feeder with a large enough seed cavity to house the weatherized Nano and XBee assembly
9. A computer (not pictured), preferably Linux or Mac-based, with Python 2.6 or higher installed to process incoming messages from the bird feeder

If you opt to use the Arduino Nano in place of the Arduino Uno, you will also need a standard A to Mini-B USB cable (not pictured) to connect the Arduino Nano to the computer. Additionally, since the Arduino Nano uses male pins for wiring connections instead of the female headers found on an Arduino Uno, you will need female jumper wires (not shown) instead of standard wires. This will allow you to more easily attach wires to the Nano's pins without having to solder the wiring connections in place.

This project is a more complex than the Water Level Notifier, and getting the XBee radios working reliably is the trickiest part. Still, it's worth the effort since you will not only have a cool twenty-first-century high-tech bird feeder, but you will also be able to reuse the XBee radio setup in several other projects. Ready to roll up your sleeves? Then let's get to it!

5.2 Building the Solution

Assembling the hardware to fit snuggly inside the feeder may require some ingenuity, especially if the bird feeder doesn't offer much space inside the seed container. Before we start cramming electronics into the feeder, we first need to make sure our components work as expected.

1. We will start with the easy part of connecting the aluminum foil capacitive sensor to the Arduino and writing a function that will send a message to the serial window (and eventually the serial port of the attached XBee radio) when the sensor is triggered.
2. Next, we will hook up the photocell to the Arduino and write the code for it to react to changes in light.
3. Then we will pair the two XBee radios and transmit these events from the XBee radio attached to the Arduino to the other XBee radio tethered to a computer via an FTDI USB cable.

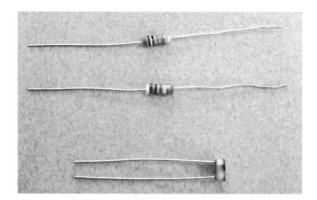

Figure 11—Tweeting Bird Feeder resistors

4. And finally, we will write a Python script that will capture the data into an SQLite database and format and transmit the messages to be posted on Twitter.

Once everything is working, we will compact the Arduino (preferably the Nano)+XBee+perch resistor+photocell assembly into a weatherized package, house it in the feeder, fill the feeder with seed, and go outside for a live field test.

5.3 The Perch Sensor

Bird feeders come in different shapes and sizes. I opted to go with a low-tech solution for determining when a bird lands on the feeder perch. While it is certainly possible to construct an elaborate pressure switch mechanism, the time and expense required to implement it seems like a lot of work just to detect when something grips the perch. Instead, we can monitor fluctuations in electrical capacitance.

Cover the feeder perch with aluminum foil, attach a wire from the foil to a resistor connected to the digital pins on the Arduino. By measuring baseline values and fluctuations detected when a bird lands on this sensor, we can establish a threshold value to determine when a landing message should be transmitted.

Building the Sensor

Building and testing the perch sensor is the easiest part of this project. Take a piece of aluminum foil, flatten it to half the size of a gum stick wrapper, and wrap it across the bird perch. Then take a 10M ohm resistor and insert one end into the Arduino's digital pin 7 and the other end in digital pin 10. Then connect a wire from the foil to the resistor lead that is connected to the Arduino's digital pin 7. For the wiring diagram, refer to Figure 12, *Wiring up a capacitive sensor*, on page 63.

Programming the Sensor

Connect the Arduino to your computer and fire up the Arduino IDE to write the sensing code. Like the Water Level Notifier project, we will write code that we will use to do the following:

1. Display the capacitive sensor's electrical values to the Arduino IDE's serial monitor window.
2. Identify the baseline electrical value of the sensor.
3. Alter the flow of current when placing a finger on the sensor.
4. Record the new value to use for our alert trigger condition.

In order to more easily and programmatically detect the electrical changes that occur when something like a finger or a bird lands on the foil, we will call upon the help of Arduino enthusiast Paul Badger. Paul wrote an Arduino library that makes measuring changes in capacitive sensors like the foil one we constructed for this project a breeze. Called the Capacitive Sensing library,[4] the library gives Arduino programmers the ability to turn two or more Arduino pins into a capacitive sensor that can be used to sense the electrical capacitance of a body. A human body is considerably larger than a bird and will therefore create a much larger deflected value. Nevertheless, a bird also has measurable electrical capacitance, and that is the threshold value we will attempt to refine in our program.

Download this library, uncompress its contents, and copy it into your Arduino's libraries folder. For more details, refer to Appendix 1, *Installing Arduino Libraries*, on page 209.

Create a new Arduino project and use the # include CapSense.h;.

Due to the difference in size and surface area touching the foil, a bird will have a very different value than that of a person. If possible, measure the value difference with a bird. Fortunately my kids have pet parakeets, and these birds were all too eager to be test subjects in exchange for the seed supplied by the feeder. My test measurements concluded that the baseline value varied between 900 and 1400, and the bird's capacitance increased that value to more than 1500. Using these values, we can use the same type

---

4. http://www.arduino.cc/playground/Main/CapSense

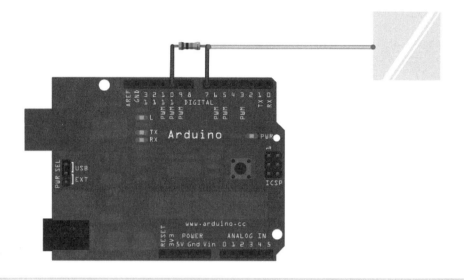

Figure 12—Wiring up a capacitive sensor

of conditional code from the Water Level Notifier project to raise and reset the bird landing and departure notifications.

We will write the code that will load the CapSense library and capture and display the capacitive values to the serial monitor window.

TweetingBirdFeeder/BirdPerchTest.pde
```
#include <CapSense.h>

#define ON_PERCH     1500
#define CAP_SENSE    30
#define ONBOARD_LED  13

CapSense foil_sensor   = CapSense(10,7); // capacitive sensor
                       // resistor bridging digital pins 10 and 7,
                       // wire attached to pin 7 side of resistor
int perch_value  = 0;
byte perch_state = 0;

void setup()
{
    // for serial window debugging
    Serial.begin(9600);

    // set pin for onboard led
    pinMode(ONBOARD_LED, OUTPUT);
}
```

```
void SendPerchAlert(int perch_value, int perch_state)
{
    digitalWrite(ONBOARD_LED, perch_state ? HIGH : LOW);
    if (perch_state)
        Serial.print("Perch arrival event, perch_value=");
    else
        Serial.print("Perch departure event, perch_value=");
    Serial.println(perch_value);
}

void loop() {
    // wait a second each loop iteration
    delay(1000);

    // poll foil perch value
    perch_value =  foil_sensor.capSense(CAP_SENSE);

    switch (perch_state)
    {
    case 0: // no bird currently on the perch
        if (perch_value >= ON_PERCH)
        {
            perch_state = 1;
            SendPerchAlert(perch_value, perch_state);
        }
        break;

    case 1: // bird currently on the perch
        if (perch_value < ON_PERCH)
        {
            perch_state = 0;
            SendPerchAlert(perch_value, perch_state);
        }
        break;
    }
}
```

Note the defined ON_PERCH value of 1500 to compare against the recorded perch_value. Due to variations in the conductivity and surface area of your foil sensor, you may need to tweak the ON_PERCH threshold value just like you did for the Water Level Notifier project so that it works best for your configuration. Also note the value of 30 assigned to the CAP_SENSE constant. This is the number of samples to poll during the capacitive measurement cycle.

Now that we have a working bird perch sensor, we need a way for the feeder to alert us when it is running low on seed. How will we do this? A photocell can help.

5.4 The Seed Sensor

A photocell measures light intensity, with higher intensity correlating with higher current and lower intensity with lower current. For a more detailed explanation and tutorial on photocells, visit Ladyada's ever-helpful website.[5] By placing the photocells at a level beneath the seeds poured into the feeder, we can detect when the seed level dips below the sensor, exposing it to more light and thereby alerting us that the feeder needs to be refilled.

Before drilling holes into the feeder for placement of the photocell, we need to write some code and test it using the same approach that we did for our homemade foil resistor.

Connect one of the photocell leads to the Arduino 5v pin and connect the other photocell lead to the Arduino analog pin 0. Then bridge a 10k ohm resistor between the Arduino analog pin 0 and the Arduino ground pin, as shown in Figure 13, *Wiring diagram for the photocell test*, on page 66. Does this electrical pattern look familiar? Yep, it's the same wiring configuration used previously with Arduino sensors in this book. This is a frequent pattern for various types of sensors that connect with the Arduino.

With the photocell connected, connect the Arduino to the computer via the USB serial cable and launch the Arduino IDE. Using the same technique used for the foil switch, monitor analog pin 0 values in the Arduino IDE's serial monitor window and capture baseline values for when the sensor is bathed in ambient, standard lighting conditions. Then cover the sensor with your finger to block incoming light. Note the difference in value.

Just as we did for the capacitive test, we will write the same type of procedures and conditional statements to test for luminosity thresholds. Indeed, you could copy and paste code from the foil test and simply change variable names and connected pin assignments to create the working program.

TweetingBirdFeeder/SeedPhotocellTest.pde
```
#define SEED              500
#define ONBOARD_LED        13
#define PHOTOCELL_SENSOR    0
int seed_value  = 0;
byte seed_state = 0;
void setup()
{
    // for serial window debugging
    Serial.begin(9600);
```

---

5. http://www.ladyada.net/learn/sensors/cds.html.

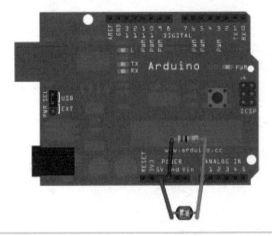

Figure 13—Wiring diagram for the photocell test

```
    // set pin for onboard led
    pinMode(ONBOARD_LED, OUTPUT);
}

void SendSeedAlert(int seed_value, int seed_state)
{
    digitalWrite(ONBOARD_LED, seed_state ? HIGH : LOW);
    if (seed_state)
        Serial.print("Refill seed, seed_value=");
    else
        Serial.print("Seed refilled, seed_value=");
    Serial.println(seed_value);
}

void loop() {
    // wait a second each loop iteration
    delay(1000);

    // poll photocell value for seeds
    seed_value = analogRead(PHOTOCELL_SENSOR);

    switch (seed_state)
    {
    case 0: // bird feeder seed filled
        if (seed_value >= SEED)
        {
            seed_state = 1;
            SendSeedAlert(seed_value, seed_state);
        }
        break;
```

```
    case 1: // bird feeder seed empty
        if (seed_value < SEED)
        {
            seed_state = 0;
            SendSeedAlert(seed_value, seed_state);
        }
        break;
    }
}
```

Measuring and assigning the defined SEED threshold value for the photocell is much easier and more reliable than the capacitive foil test we did earlier. While you can use your finger to cover up the photocell and measure the value change, it is more authentic to test with real seed. If you don't want to drill holes in your bird feeder to set the photocell just yet, use a paper cup and place the photocell toward the bottom of the cup.

Similar to the calibration procedure we used in the Water Level Notifier project, add these lines after the seed_value = analogRead(PHOTOCELL_SENSOR); request in the sketch's main program loop:

```
Serial.print("seed_value=");
Serial.println(seed_value);
```

Record the seed_value starting value, then fill the cup with seed and measure the new value. Use these values to set the starting and threshold values for the photocell.

If you do not see any change, check your wiring and try again. In my tests with the photocell, the baseline value fluctuated between 450 and 550. It reported below 100 when my finger covered the sensor. Use whatever upper and lower limit values you recorded with your tests, keeping in mind that they will need to be recalibrated once the sensor is mounted inside the feeder.

Now that monitoring is working for both the perch and light sensors, we need a way to communicate when those sensor thresholds have been exceeded. It's not very practical to run an Ethernet cable from an indoor network hub to an outdoor tree limb. Not to mention that trying to fit a bulky Arduino+Ethernet shield assembly into the confined space of a bird feeder would be a challenge. We will use the convenience of low-power wireless communication to transmit these sensor notifications to an indoor computer. Then we will use that computer's faster processing and larger storage capacity to analyze and act upon the data received.

5.5 Going Wireless

Although more ubiquitous 802.11b/g Wi-Fi shields exist for the Arduino (such as Sparkfun's WiFly Shield), the most prevalent means of Arduino wireless communication is via XBee radios. The initial outlay for a set of XBee devices can be a bit pricey. This is because in addition to the XBee radios, you also need an FTDI USB cable to connect one of the radios to a computer to act as a wireless serial port.

The other XBee is typically connected to an Arduino. And to make it easier to make these connections, additional kits are available to mount the XBee radios to an assembly that better exposes the connection pins while also displaying data transfers via onboard LEDs. Such visual indicators can be quite helpful when debugging or troubleshooting a paired XBee connection. Nevertheless, the fact that XBees offer a low-power and long-range (up to 150 feet) solution coupled with easy connectivity and data transfer protocol make them an ideal wireless technology for our project needs.

In order to more easily interface with the XBee radios, Adafruit has designed an adapter kit that "doesn't suck," but it does require you to solder a few small components onto the adapter board. Assemble the adapter following the instructions posted on Ladyada's website.[6]

Once assembled, configuring and pairing the XBee radios isn't too difficult, though one of the more helpful utilities used to configure them only runs on Windows.

Following the instructions for the XBee point-to-point sample program available on Ladyada's web page,[7] attach the power, ground (Gnd), receive (RX), and transmit (TX) pins of one of the adapter-mounted XBees to the Arduino's 5V, Gnd, digital 2, and digital 3 pins. Plug the Arduino into your computer, upload the test program, open up the Arduino serial window, and make sure it is set to a rate of 9600 baud. Leave the Arduino plugged in and attach the other paired XBee to your computer via the FTDI USB cable. Open up a serial terminal session: Hyperterminal in Windows, the screen on a Mac, or various serial communication programs such as Minicom for Linux.[8]

Once you have established a serial connection, type in a few characters in your serial application's input screen. If your XBee radios are properly configured, you will see those typed characters appear in the Arduino serial window.

---

6. http://www.ladyada.net/make/xbee/
7. http://ladyada.net/make/xbee/point2point.html
8. http://alioth.debian.org/projects/minicom/

> **Using the screen Utility**
>
> A helpful serial monitoring utility for *nix-based platforms like Mac OS X and Linux is the screen application. To use screen in OS X, determine the serial port that the FTDI USB cable is using by launching the Arduino IDE. Select Serial Port from the Arduino IDE Tools menu to identify the serial port assigned to it.
>
> In my case, the device name of the FTDI USB-to-XBee adapter connection was /dev/tty.usbserial-A6003SHc, though yours may be different, depending on other devices connected to your computer. Open up the Terminal application and type screen /dev/tty.YOURDEVICE 9600. This will open the serial port and allow you to enter and receive characters at a communication rate of 9600 baud. To gracefully exit the screen utility, press Control-A followed by Control-\.

If you connected the XBees to the Arduino and the FTDI cable using XBee adapters, you should also see the adapter's green transmit and red receive LEDs blinking as the characters are wirelessly transmitted from one XBee radio to the other.

If you do not see the characters being displayed on the receiving window, review the XBee radio wiring to the appropriate Arduino ports. You can also swap the radios to verify both are recognized using the FTDI USB connection. Type an AT in your serial application's terminal window and verify that it returns an OK acknowledgment.

If the XBees still don't seem to be communicating with one another, contact the retailer you purchased them from for further assistance.

With the XBee radios successfully paired, we will reconnect both the photocell and foil resistor to the XBee-outfitted Arduino and combine the code that will receive the sensor's trigger condition events. For the complete wiring diagram, refer to Figure 14, *Tweeting bird feeder with sensors and XBee radio attached to Arduino*, on page 70.

Consider using a breadboard, or solder the resistors along with the wires being connected to the Arduino pins. If you opt for a breadboard for testing purposes, keep in mind that it probably won't fit in the feeder, so be prepared to solder the wiring permanently in place after the tests prove successful. And depending on the orientation of the Arduino Uno or Nano as it is positioned into the bird feeder, you may need to use straight header pins on the XBee adapter board instead of the default right-angled pins that accompany the kit. The goal is to make sure everything fits inside the feeder in a secure and serviceable way. And recall that unlike the female headers on an Arduino Uno, the Arduino Nano uses male pins instead. As such, you will need to use

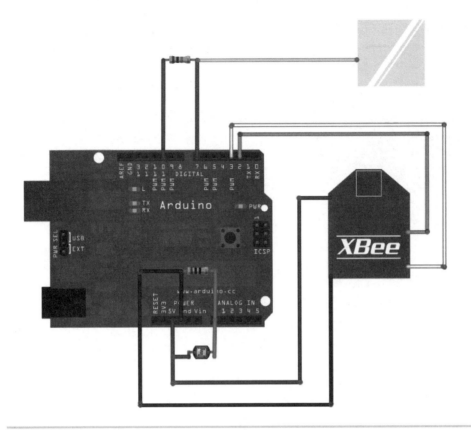

Figure 14—Tweeting bird feeder with sensors and XBee radio attached to Arduino

female jumper wires to better accommodate the connections to the Nano's male pins.

Finishing the Sketch

We are going to poll both the foil and photocell sensors for activity, first to the Arduino IDE serial window, then—with one small replacement in our code—to the XBees. Essentially, we are simply going to combine the threshold code for the foil and photocell that we tested earlier while sending threshold alerts to the serial port of the XBee radio. Adding this logic to what we already wrote to test the perch and seed status, we can complete the sketch for the project.

TweetingBirdFeeder/TweetingBirdFeeder.pde
```
#include <CapSense.h>;
#include <NewSoftSerial.h>
```

```
#define ON_PERCH        1500
#define SEED             500
#define CAP_SENSE         30
#define ONBOARD_LED       13
#define PHOTOCELL_SENSOR   0
// Set the XBee serial transmit/receive digital pins
NewSoftSerial XBeeSerial = NewSoftSerial(2, 3);
CapSense foil_sensor     = CapSense(10,7); // capacitive sensor
                         // resistor bridging digital pins 10 and 7,
                         // wire attached to pin 7 side of resistor
int perch_value  = 0;
byte perch_state = 0;
int seed_value   = 0;
byte seed_state  = 0;

void setup()
{
    // for serial window debugging
    Serial.begin(9600);

    // for XBee transmission
    XBeeSerial.begin(9600);

    // set pin for onboard led
    pinMode(ONBOARD_LED, OUTPUT);
}

void SendPerchAlert(int perch_value, int perch_state)
{
    digitalWrite(ONBOARD_LED, perch_state ? HIGH : LOW);
    if (perch_state)
    {
        XBeeSerial.println("arrived");
        Serial.print("Perch arrival event, perch_value=");
        }
    else
    {
        XBeeSerial.println("departed");
        Serial.print("Perch departure event, perch_value=");
    }
    Serial.println(perch_value);
}

void SendSeedAlert(int seed_value, int seed_state)
{
    digitalWrite(ONBOARD_LED, seed_state ? HIGH : LOW);
    if (seed_state)
    {
        XBeeSerial.println("refill");
        Serial.print("Refill seed, seed_value=");
```

```
        }
        else
        {
            XBeeSerial.println("seedOK");
            Serial.print("Seed refilled, seed_value=");
        }
        Serial.println(seed_value);
}

void loop() {
    // wait a second each loop iteration
    delay(1000);

    // poll foil perch value
    perch_value =  foil_sensor.capSense(CAP_SENSE);

    // poll photocell value for seeds
    seed_value =  analogRead(PHOTOCELL_SENSOR);

    switch (perch_state)
    {
    case 0: // no bird currently on the perch
        if (perch_value >= ON_PERCH)
        {
            perch_state = 1;
            SendPerchAlert(perch_value, perch_state);
        }
        break;

    case 1: // bird currently on the perch
        if (perch_value < ON_PERCH)
        {
            perch_state = 0;
            SendPerchAlert(perch_value, perch_state);
        }
        break;
    }

    switch (seed_state)
    {
    case 0: // bird feeder seed filled
        if (seed_value >= SEED)
        {
            seed_state = 1;
            SendSeedAlert(seed_value, seed_state);
        }
        break;

    case 1: // bird feeder seed empty
        if (seed_value < SEED)
```

```
        {
            seed_state = 0;
            SendSeedAlert(seed_value, seed_state);
        }
        break;
    }
}
```

Note the references to the capacitive sensing and new software serial libraries at the beginning of the sketch. We initialize the variables that will use references from those libraries along with our threshold variables. Then we set up the connections to the serial window and XBee radio along with the Arduino onboard LED on pin 13. Once initialized, the program simply runs a loop and waits for the perch or seed thresholds to be exceeded or reset. If a change condition is detected, the running sketch will transmit these trigger messages to the Arduino IDE's serial window as well as to the XBee radio.

With both the foil and photocells correctly reporting their values, redirect serial output from the Arduino IDE serial window to the XBee attached to the Arduino. Open up the serial application window to the FTDI-connected XBee, and if all goes well, the data you saw being displayed in the Arduino IDE serial window should now be showing up in the FTDI-connected serial application window on your computer. Isn't wireless communication cool?

At this point, the hardware for our project is connected and tested and may look similar to the bird feeder components I assembled in Figure 15, *A pet bird can help out with the testing and debugging of threshold values that trigger perch landing and departure events*, on page 74.

But before we start packing the hardware into the bird feeder, we need to create one more crucial component. Using the Python language, we can write a short program that will listen for bird landings, poll for seed status, and post notable changes to Twitter. Let's go write some code.

5.6 Tweeting with Python

A number of different languages capable of accomplishing the tasks of monitoring and interpreting incoming serial console messages and transmitting outbound messages to the serial port exist. There are also a number of Twitter libraries for various programming languages.

Python was chosen for this and several other scripts in the book due to the language's easy-to-follow syntax, its default inclusion in Linux and Mac OS X operating systems, and its "batteries included" approach to bundling a

Figure 15—A pet bird can help out with the testing and debugging of threshold values that trigger perch landing and departure events.

number of relevant libraries (such as SQLite) in its base distribution. To learn more about programming in Python, check out *Learning Python* [LA03].

For this project, the basic functionality we need the tweetingbirdfeeder.py Python script to accomplish is this:

1. Record the date and time when a bird lands and departs from the perch in the birdfeeding table in the tweetingbirdfeeder database.

2. Record the date and time when seed levels are depleted and replenished in the seedstatus table, which is also part of the tweetingbirdfeeder database.

3. Listen for inbound and transmit outbound serial messages via the FTDI cable-connected XBee radio and respond to events by storing the date and time of their occurrence and condition.

4. Connect to Twitter via OAuth authentication and post changes in bird feeding and seed level status.

The only additional Python libraries that need to be installed for this project are pyserial and python-twitter.

Beyond relaying tweets to a designated Twitter account, it would be helpful to visualize trends in the data we will be tweeting, such as the frequency and date/time of bird landings and the average time between seed refills. We can then see how these trends map out over an hour, a day, a month, and a year. To do this, we will need to capture the data in a structured format.

Configure the Database

Since Python 2.5 and higher supports the SQLite database out of the box, and because our data needs don't require an overengineered standalone database server, SQLite is the ideal choice for the job. While we could have dumped these values to a plain-text comma separated value (CSV) file, organizing the data into a structured SQLite file affords us two benefits: First, we will be better prepared for future data analysis queries. Second, we will have greater flexibility to capture and manage other types of data events later on simply by adding new columns to the table.

In order to create a database in sqlite3 file format, we can use the sqlite3 command-line tool. This tool is already installed on Mac OS X. On Linux, it will most likely need to be retrieved from the distribution's repository. In the case of Debian-based distributions like Ubuntu, issuing sudo apt-get install sqlite3 libsqlite3-dev should install the application. Windows users will need to download the sqlite3.exe utility from the SQLite website.[9]

Once installed, type sqlite3 within a terminal window. This will display something like the following:

```
SQLite version 3.7.6
Enter ".help" for instructions
Enter SQL statements terminated with a ";"
sqlite>
```

Your installation of SQLite may report a different version number.

Next, we will enter the SQL statement to create our new database. To do so, exit the sqlite command shell by typing .q followed by a carriage return at the sqlite> prompt. Then relaunch the sqlite3 tool, followed by the name of the database to be opened.

For this project, we will call the database tweetingbirdfeeder, with the filename tweetingbirdfeeder.sqlite. Because this database file does not yet exist, SQLite will

---

9. http://www.sqlite.org/download.html

automatically create the file for us. The database file will be created from the same directory that you launched the sqlite3 tool from. For example, if you launched sqlite3 from your home directory, the database file will be created there.

We will create a new table in the tweetingbirdfeeder.sqlite database that we will call birdfeeding with the following structure:

Column Name	Data Type	Primary Key?	Autoinc?	Allow Null?	Unique?
id	INTEGER	YES	YES	NO	YES
time	DATETIME	NO	NO	NO	NO
event	TEXT	NO	NO	NO	NO

We can create this table by submitting the following SQL statement to the sqlite command-line tool:

```
[~]$ sqlite3 tweetingbirdfeeder.sqlite
SQLite version 3.7.6
Enter ".help" for instructions
Enter SQL statements terminated with a ";"
sqlite> CREATE TABLE "birdfeeding" ("id" INTEGER PRIMARY KEY NOT NULL UNIQUE,
"time" DATETIME NOT NULL,"event" TEXT NOT NULL);
```

With the birdfeeding table established, we need another table, one that has a similar structure in the same database and is called seedstatus:

Column Name	Data Type	Primary Key?	Autoinc?	Allow Null?	Unique?
id	INTEGER	YES	YES	NO	YES
time	DATETIME	NO	NO	NO	NO
event	TEXT	NO	NO	NO	NO

Just like the birdfeeding table, submitting the following SQL statement to the sqlite command-line tool will generate the desired structure for the seedstatus table:

```
[~]$ sqlite3 tweetingbirdfeeder.sqlite
SQLite version 3.7.6
Enter ".help" for instructions
Enter SQL statements terminated with a ";"
sqlite> CREATE TABLE "seedstatus" ("id" INTEGER PRIMARY KEY  NOT NULL,
"time" DATETIME NOT NULL ,"event" TEXT NOT NULL );
```

Now that the database has been defined, we can work on the code to import the database and Serial and Twitter libraries, then listen for serial events being generated and timestamp and store these events to the appropriate database table.

> **SQLite Manager**
>
> While the SQLite command-line tools provide all you need to create and manage SQLite databases, it's sometimes easier to work with a graphic user interface. This is especially applicable if you need to scroll through rows of data in a single window. Several open source GUI-based SQLite database explorer-style apps exist. If you are a Mozilla Firefox web browser user, I recommend using the cross-platform SQLite Manager Firefox plug-in.[a]
>
> Installing the plug-in is easy. From the Firefox's Tools→Add-ons menu selection, search for SQLite Manager and click the Install button. Open SQLite Manager from Firefox's Tools→SQLite Manager menu option. Creating a new database is as simple as clicking on the New Database icon in the SQLite Manager window toolbar. Saving and opening SQLite database files is just as easy.
>
> ---
> a. https://addons.mozilla.org/en-US/firefox/addon/sqlite-manager/

We'll conclude the event capture by posting a tweet of the situation. But before you can programmatically tweet to Twitter, you need to create a Twitter account and sign up for a Twitter API key and related OAuth credentials. Let's go get ourselves an API key.

Twitter API Credentials

Before sending tweets to Twitter, you need a Twitter account to send them to. And before you send tweets to Twitter programmatically via a language or library that supports OAuth,[11] you need to create an application identifier for the intended Twitter account. While you could use an existing Twitter account, I prefer creating new accounts whenever a new project demands it. That way, followers of my existing account are not accosted by tweets of my latest experiments. It also offers a way to share your application's tweets selectively. With these considerations in mind, create a new account and application ID specifically for the bird feeder project.

Using your new Twitter account credentials, visit dev.twitter.com and select the "Register an app" option. On the New Twitter Application page, enter a unique name for your application, a description at least ten characters long, and a valid website for the app. Enter a temporary one if you don't have a permanent website from which you will offer your application for download. Then select Client under Application Type and select Read & Write under Default Access type. You can upload a custom icon if you like, but it's not required. Then enter the CAPTCHA validation and click the Register Application button at

---
11. http://oauth.net/

the bottom of the screen. Read and accept the Twitter API Terms of Service to proceed.

Once your request has been approved, a unique API Key, OAuth Consumer key, and Consumer secret will be generated. Click the My Access Token menu item on the right side of the page to access your application's all important Access Token (oauth_token) and super-secret Access Token Secret (oauth_token_secret). Copy these unique codes and store them in a safe, secure file. You will need these values to programmatically interact with your new Twitter account. Remember to keep these values a secret! You don't want any unscrupulous individuals getting hold of your secret token and using it to spam your friends and raise the ire of the Twitter community.

With a Twitter account and a valid application API key in hand, you can use these credentials in our Python-based Tweeting Bird Feeder application.

The Python-Twitter Library

Even though we have API access to Twitter, we still need to talk to Twitter from Python. We can do so with help a little help from the Python-Twitter library.[12] To install both the Pyserial and Python-Twitter libraries, download the latest version and execute the standard sudo python setup.py install command. If you're installing this library on Mac OS X 10.6 (Snow Leopard) or higher, the easy_install Python setup tool is preinstalled. However, due to quirks in the 64-bit libraries, you will need to precede the command with an i386 architecture flag to install the Python-Twitter library without errors. The complete command for this is sudo env ARCHFLAGS="-arch i386" easy_install python-twitter.

At last, we have all the accounts configured and dependencies installed, and we can complete the project by writing the Python script that will listen for messages on the receiving XBee radio serial port, timestamp the messages, save them in a database, and post the message to Twitter. Let's write the Python script that will codify this process.

TweetingBirdFeeder/tweetingbirdfeeder.py
```python
# import DateTime, Serial port, SQLite3 and Twitter python libraries
from datetime import datetime
import serial
import sqlite3
import twitter

# import the os module to clear the terminal window at start of the program
# windows uses "cls" while Linux and OS X use the "clear" command
import os
```

---

12. http://code.google.com/p/python-twitter/

```python
    if sys.platform == "win32":
        os.system("cls")
else:
    os.system("clear")

# Connect to the serial port, replacing YOUR_SERIAL_DEVICE with the
# name of the serial port of the FTDI cable-attached XBee adapter
XBeePort = serial.Serial('/dev/tty.YOUR_SERIAL_DEVICE', \
                        baudrate = 9600, timeout = 1)

# Connect to SQLite database file
sqlconnection = sqlite3.connect("tweetingbirdfeeder.sqlite3")

# create database cursor
sqlcursor = sqlconnection.cursor()

# Initialize Twitter API object
api = twitter.Api('Your_OAuth_Consumer_Key', 'Your_OAuth_Consumer_Secret', \
        'Your_OAuth_Access_Token', 'Your_OAuth_Access_Token_Secret')

def transmit(msg):
    # Get current date and time and format it accordingly
    timestamp = datetime.now().strftime("%Y-%m-%d %H:%M:%S")

    # Determine message and assign response parameters
    if msg == "arrived":
        tweet = "A bird has landed on the perch!"
        table = "birdfeeding"
    if msg == "departed":
        tweet = "A bird has left the perch!"
        table = "birdfeeding"
    if msg == "refill":
        tweet = "The feeder is empty."
        table = "seedstatus"
    if msg == "seedOK":
        tweet = "The feeder has been refilled with seed."
        table = "seedstatus"

    print "%s - %s" % (timestamp.strftime("%Y-%m-%d %H:%M:%S"), tweet)

    # Store the event in the SQLite database file
    try:
        sqlstatement = "INSERT INTO %s (id, time, event) \
        VALUES(NULL, \"%s\", \"%s\")" % (table, timestamp, msg)
        sqlcursor.execute(sqlstatement)
        sqlconnection.commit()
    except:
        print "Could not store event to the database."
        pass
```

```python
    # Post message to Twitter
    try:
        status = api.PostUpdate(msg)
    except:
        print "Could not post Tweet to Twitter"
        pass

# Main program loop
try:
    while 1:
        # listen for inbound characters from the feeder-mounted XBee radio
        message = XBeePort.readline()

        # Depending on the type of message is received,
        # log and tweet it accordingly
        if "arrived" in message:
            transmit("arrived")

        if "departed" in message:
            transmit("departed")

        if "refill" in message:
            transmit("refill")

        if "seedOK" in message:
            transmit("seedOK")

except KeyboardInterrupt:
    # Exit the program when the Control-C keyboard interrupt been detected
    print("\nQuitting the Tweeting Bird Feeder Listener Program.\n")
    sqlcursor.close()
    pass
```

Once the necessary datetime, serial, sqlite, and twitter libraries are loaded, we clear the terminal window (sending a cls for Windows and a clear for any other operating system), connect to the receiving XBee radio serial port (the XBee that is attached to the computer via the FTDI cable). Then we connect to the tweetingbirdfeeder.sqlite3 database file we created earlier and run an infinite while loop until the Control-C keyboard combination is triggered so we can gracefully exit the program. If the attached XBee radio receives a message it recognizes, we call the def transmit(msg) function that parses the msg variable, adds descriptive text to the event, saves the message to the database, and posts it to Twitter.

With the Arduino running and the XBee radios paired and powered, test the threshold detections by touching the perch sensor and photocell enough times to trigger several event transmissions. Assuming no errors were reported in the terminal window of the running script, open the tweetingbirdfeeder.sqlite3 file in the SQLite Manager's Browse and Search window and verify that entries for both sensors were timestamped when the related events were triggered. If everything checks out, log into the Twitter account that you used to post the events and verify that the events appear on the timeline.

We're almost done. Just a few more hardware assembly steps remain.

5.7 Putting It All Together

In order to make the project fully functional, we need to package up the Arduino+XBee hardware assembly inside a weatherized, protected layer within the bird feeder, mount the photocell near the base of the feeder, fill the feeder with seed, attach the Arduino+XBee hardware to a power source, and place the feeder outdoors but within range of the paired XBee radio attached to the computer.

Unless you live in an area with little rainfall, you will need to protect the electrical assembly from water damage. I have found double bagging the components in a small plastic freezer bag does a sufficient job of weatherproofing the Arduino+XBee. However, unless you plan on powering the electronics with a 9V battery that can be contained within the bundle (it might be good for short data collection sessions but won't last very long before its charge is exhausted), you will need to account for an external cord to attach to the Arduino so that continuous power can be delivered.

Cutting a small opening in the bag to allow the cable to enter works, but doing so exposes the insides to potential moisture condensation. To minimize this risk, tightly wrap the freezer bag and cable exit point with a continuous sheet of plastic wrap, climbing high enough up the power cord to ensure a good seal that won't slip or loosen with weather changes.

Using a weatherized power cord (such as those sold for outdoor decorative lighting purposes) may be less expensive and easier to test in the short term. However, environmentally conscious individuals may prefer instead to spend a bit more money up front for a longer, more sustainable energy alternative in the form of a photovoltaic power supply.

When searching for an adequate, portable solar power solution, make sure it can deliver 5 volts, is built for rugged durability, and has a built-in rechargeable battery when backup power is needed. Products like the Solio Bolt provide a relatively inexpensive solution for short-term measurements.[13]

---

13. http://www.solio.com/chargers/

If you prefer photovoltaic solutions that offer greater internal battery charging capacities and voltage, be prepared to spend a bit more for the added capabilities. Companies like Sunforce Products offer a variety of solar backup power maintainers, trickle chargers, and controllers designed to take on greater loads.[14]

You should mount the solar panel far enough away from the feeder to gain maximum sun exposure. If possible, mount the panel at a ninety-degree angle to the sun for optimal energy capture. Depending on your location and average level of daylight intensity, you may need to seek alternatives such as consumer-grade wind turbine chargers or even pedal-powered dynamos.

We have accomplished quite a number of new objectives this project, from using photocell and homemade sensors and learning how to pair and wirelessly communicate between XBee-attached hardware to writing a script that records structured data, responds to events, and submits posts to Twitter via Twitter's API. We have also taken into account standalone Arduino+XBee radio energy requirements and ways to adequately shield our electronics from environmental changes.

We will reuse these valuable lessons with some of the other projects in the ensuing chapters.

5.8 Next Steps

The variety of home automation projects using capacitive and photocell sensor notifications is expansive. Here are just a few ideas to consider pursuing:

- Place an XBee+photocell-equipped Arduino, powered by a rechargeable battery pack, in your refrigerator or freezer to detect how often and how long the doors are left open. Use this data to calculate energy expended each month as a result of these encounters. If such expenditures are excessive, broadcast emails and/or tweets to cohabitants reminding them of their growing carbon footprint.

- If groping for a light switch is a frustratingly frequent chore, tweak the capacitive sensor configuration to be used to turn on basement or garage lights by touching foil tacked to a wall or to a table-mounted surface.

- Use the variable analog readings of a photocell to measure day-night cycles and sunny-overcast recordings and map these to seasonal gardening data. Did planting certain species of flowers, fruits, or vegetables before a certain

---

14. http://www.sunforceproducts.com/results.php?CAT_ID=1

time help or hinder their growth? How many days were the plants bathed in full sunlight compared to overcast or inclement skies?

In addition to these suggestions, there is still plenty of data analysis that can be done with the collected bird feeder data. Use a Python graphing library like CairoPlot to help visualize the average duration of and time in between bird visits.[15] How quickly was seed consumed? How much of an effect did weather have on feeding times and durations? Does changing the type of seed alter the duration and frequency of visits?

Consider sharing your tweets with other bird enthusiasts online to expand a social network of others collecting and sharing their bird feeder data. Collectively compare patterns across various regions and geographies to infer population trends, environmental impacts, migration cycles, and other factors to better understand the habits of our feathered friends.

---

15. http://cairoplot.sourceforge.net/

CHAPTER 6

Package Delivery Detector

There is nothing like that surge of anticipation when coming home after a long day to discover a package delivery resting outside the front door. Even if it was a package you were expecting and tracked online with every departure hub, seeing that parcel safely awaiting your arrival can sometimes feel like receiving a birthday gift.

Wouldn't it be even more comforting to know the very moment when your package arrived versus waiting up to hours later for an email from the courier confirming the delivery? (See Figure 16, *Receive an email from your home whenever a package arrives*, on page 86) What if the driver accidentally delivered the package to the wrong location? Say goodbye to those worries. The Package Delivery Detector will send you an email when a package is left at your doorstep. You can further filter the notification by delaying the message until the shipper confirms delivery via its web services.

This project combines the components we used in Chapter 5, *Tweeting Bird Feeder*, on page 57, with a similar alert monitoring mechanism used in Chapter 4, *Electric Guard Dog*, on page 43. Instead of using a PIR, the Package Delivery Detector will sense deliveries with a force sensitive resistor. When a box approximating the weight of a small package (or about half a kilogram) is dropped on top of a delivery pad containing a force sensitive resistor, the sensor sends a notification via an XBee-connected serial port. In turn, this notification processes a Python script that logs the delivery and sends an email notification. Optionally, the delivery notification can wait an hour and confirm delivery with the courier's website before transmitting the verified message.

Figure 16—Receive an email from your home whenever a package arrives.

6.1 What You Need

Most of what you need to build this project has been used by other projects in the book, with the exception of a force sensitive resistor (sometimes erroneously referred to as a pressure sensor). Take a look at the complete list of the project's components (refer to the photo in Figure 17, *Package Delivery Detector parts*, on page 87):

1. An Arduino Diecimila, Nano, or Uno
2. Paired XBee radios and accompanying FTDI cable
3. One 10k ohm resistor
4. A force sensitive resistor,[1] as shown in the photo in Figure 18, *Package Delivery Detector resistors*, on page 88, along with a 10k ohm resistor[2]

---

1. http://www.adafruit.com/products/166
2. Sparkfun offers a square resistor that provides even greater surface area:http://www.sparkfun.com/products/9376.

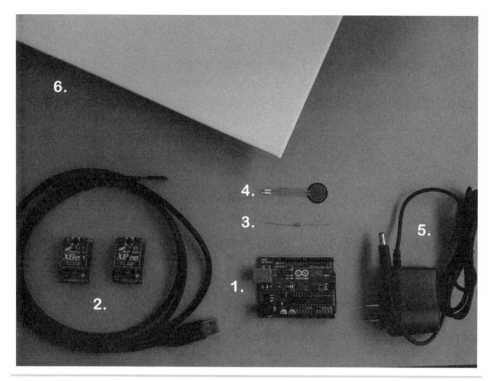

Figure 17—Package Delivery Detector parts

5. A 9-volt power supply to power the Arduino once untethered from the USB development cable
6. Two wood or plastic square tiles, preferably connected together via a wedge on one side
7. A computer (not pictured), preferably Linux or Mac-based, with Python 2.6 or higher installed to process incoming messages and interrogate popular logistics firms' web services

Depending on the type of Arduino that you opt to use for this project, you will also need a standard A to B or A to Mini-B USB cable to connect the Arduino to the computer.

If you have built the other projects in this book up to this point, the Package Delivery Detector is going to be a relatively easy project to construct. This is because it's essentially a variation of the *Tweeting Bird Feeder*. Instead of a light sensor, this project will use a strategically placed force sensitive resistor. And we will enhance the bird feeder Python script to poll a popular courier delivery service. Doing so will help certify the authenticity of the delivery. Let's get started!

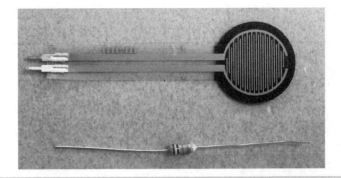

Figure 18—Package Delivery Detector resistors

6.2 Building the Solution

The hardware construction for this project closely matches the sensor approach we built for the Tweeting Bird Feeder project. We will once again call upon the Python language to script the server-centric interactions, but this script will also utilize several custom packages to interrogate popular web services offered by US-based courier companies like Federal Express (FedEx) and United Parcel Service (UPS). More specifically, we will:

1. Attach the force sensitive resistor to an available analog pin on the Arduino and identify a threshold value when weight is applied to the resistor.
2. Attach the XBee radio to the Arduino and transmit a message to another XBee radio attached to a computer when the threshold value of the force sensitive resistor has been exceeded.
3. When a threshold event has been received, pause execution for ten minutes to give the courier's tracking systems an opportunity to update its records. Then iterate through a database table of known FedEx and UPS in-transit package tracking numbers. Query these numbers with FedEx and UPS web services to determine a delivery confirmation match.
4. If a match is identified, update the tracking number database table with a delivery confirmation and date/time stamp.
5. Send an email via Google's Gmail SMTP mail gateway indicating the time of the threshold event and any packages that match the delivered status query. If no tracking number is matched, indicate such in the body of the email.

Now let's begin by first assembling the package sensor hardware components, followed by the software to drive it.

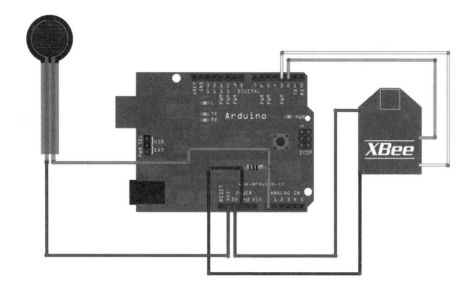

Figure 19—The Package Delivery Detector wiring diagram

6.3 Hardware Assembly

If you already constructed the Tweeting Bird Feeder project, you know how to connect the XBee and a sensor to the Arduino. If you need a refresher, refer to Section 5.5, *Going Wireless*, on page 68. Instead of attaching the light sensor with the inline 10k ohm resistor to the Arduino's analog pin, we are going to swap out the light sensor with a force sensitive resistor. See Figure 19, *The Package Delivery Detector wiring diagram*, on page 89. Wire one of the force sensitive resistor's leads to the 3.3v power source. Connect the other to analog pin 0. Then bridge the analog 0 wire to ground with a 10k ohm resistor.

The XBee radio attaches in the same way it did in the Tweeting Bird Feeder project. Namely, connect the XBee's power lead to the Arduino's 5.5v output pin. Wire the XBee's ground lead to the other available ground pin on the Arduino. Then connect the XBee's receive lead to the Arduino's digital pin 2 and the XBee's transmit lead to the Arduino's digital pin 3. Once everything is connected, it should look something like Figure 20, *A Package Delivery Detector*, on page 90. Attach the USB cable from the computer to the Arduino's USB port to power up the Arduino so we can write, run, and debug the package sensor sketch.

Figure 20—A Package Delivery Detector

6.4 Writing the Code

There are two code components in this project. The first is the sketch that monitors when something has been placed on the force sensitive resistor that is heavy enough to exceed the normal threshold. When that occurs, broadcast the event along with the value of the force resistor resistor via the XBee soft serial port.

The second component is a Python script that waits for a threshold event from the force sensitive resistor monitor. If the threshold has been exceeded, record the value of the resistor along with the date and time of the event to an SQLite database. If the database contains a list of known tracking numbers, iterate through these numbers and interrogate FedEx and UPS web services for a match. Then send an email containing information about the delivery event as well as the courier's package delivery confirmation, if available.

6.5 The Package Delivery Sketch

The code for this sketch is a variation of the code we wrote for the Tweeting Bird Feeder. One of the neat things about Arduino-centric projects is once you have written a sketch for a sensor or an actuator, the logic and syntax for the sketch can frequently be reused. After all, the basic principles of sensors and actuators are essentially the same—only the type of sensor or motor hardware and values have changed.

Since we already discussed the majority of the code in this sketch from the Tweeting Bird Feeder chapter, we're not going to spend a lot of time reviewing it. But the one variable worth mentioning is force_value. Like the other sensors we used in the other projects, you will need to calibrate the force sensitive resistor for your configuration due to the variety of force sensitive resistors available, the type of wiring and voltages used, and the way the sensor is wedged in place.

PackageDeliveryDetector/PackageDeliveryDetector.pde
```
#include <NewSoftSerial.h>

#define FORCE_THRESHOLD 400
#define ONBOARD_LED     13
#define FORCE_SENSOR     0

// Set the XBee serial transmit/receive digital pins
NewSoftSerial XBeeSerial = NewSoftSerial(2, 3);
int force_value  = 0;
byte force_state = 0;

void setup()
{
    // for serial window debugging
    Serial.begin(9600);

    // for XBee transmission
    XBeeSerial.begin(9600);

    // set pin for onboard led
    pinMode(ONBOARD_LED, OUTPUT);
}
void SendDeliveryAlert(int force_value, int force_state)
{
    digitalWrite(ONBOARD_LED, force_state ? HIGH : LOW);
    if (force_state)
        Serial.print("Package delivered, force_value=");
    else
        Serial.print("Package removed, force_value=");
    Serial.println(force_value);
    XBeeSerial.println(force_value);
}
void loop()
{
    // wait a second each loop iteration
    delay(1000);

    // poll FLEX_SENSOR voltage
    force_value = analogRead(FORCE_SENSOR);
```

```
    switch (force_state)
    {
    case 0: // check if package was delivered
        if (force_value >= FORCE_THRESHOLD)
        {
            force_state = 1;
            SendDeliveryAlert(force_value, force_state);
        }
        break;

    case 1: // check if package was removed
        if (force_value < FORCE_THRESHOLD)
        {
            force_state = 0;
            SendDeliveryAlert(force_value, force_state);
        }
        break;
    }
}
```

Even though the basic sketch is written, we still need to test the sketch on the Arduino and verify that the force sensitive resistor reacts appropriately to weight change events. We also need to ensure that the XBee radios are paired with each other and are passing the force values and weight messages being detected on the Arduino's analog pin 0.

6.6 Testing the Delivery Sketch

Install and run the sketch on the Arduino, open the Arduino IDE serial window, and pinch the force sensitive resistor between your thumb and forefinger. A delivery detection should register on the serial window. Release the force sensitive resistor and wait a few seconds. The serial window should report a value less than 400, followed by an Empty alert. If you don't see these messages, check your sensor wiring. You may also need to increase or decrease the force_sensor_value threshold value condition to address any jitter or unexpected fluctuations in the analog readings of the sensor.

Next, make sure that the XBee radios are connected and communicating with each other. Use the screen command that was mentioned in Chapter 5, *Tweeting Bird Feeder*, on page 57, to observe the inbound messages from the force sensitive resistor when it is squeezed. The information being transmitted should be the same as what is being displayed in the Arduino IDE serial window. Once everything checks out, we can write a Python script that will listen for inbound XBee messages via the FTDI cable created–serial port and act on them accordingly.

6.7 The Delivery Processor

Once again, we're going to borrow from the code we wrote in Chapter 5, *Tweeting Bird Feeder*, on page 57. We're going to copy the serial monitoring and SQLite database connectivity instructions and enhance the script with additional functionality. First, we're going to add the ability to scan a database containing known tracking numbers of packages in transit. When a delivery notification is received via the XBee/FTDI serial port connection, we will wait several minutes before iterating over the tracking numbers to determine which package was delivered. This delay can sometimes take an hour or more with some couriers. In the case of the US Post Office, it can take up to a day, making lookups on USPS deliveries impractical for our more immediate needs.

After scanning and polling the tracking numbers with the appropriate courier's web services, we will add any courier delivery validation information to our delivery notification email message. If an error occurs with the tracking number lookup or if there were no confirmed deliveries of the tracking numbers we iterated upon, we will say that in the body of the message as well.

Finally, we will use Gmail to send our message to the intended recipient. If you don't have a Gmail account, you will need to create one for this project. Alternatively, if you have SMTP outbound mail access via a different server, you're welcome to substitute Gmail's SMTP gateway with your own. Before we can write any Python code, though, we will need to create our database structures to store delivery events, tracking numbers, package descriptions, and confirmations.

6.8 Creating the Delivery Database

We need to create two tables for this project. The first will store both the history of force sensitive resistor triggers that occur when packages are delivered and removed as well as the value of the exceeded threshold value. The second table will store known tracking numbers of inbound package deliveries and a date delivery field that will contain the time and date of when the delivery was confirmed by the courier. We've done something like this before in Chapter 5, *Tweeting Bird Feeder*, on page 57, so we'll apply that same approach to the creation of the package delivery database.

We will first create the database file using the `sqlite3` tool, followed by the creation of the two tables within the `packagedelivery` database. Recall that we need to capture the force sensitive resistor's trigger actions and record the time and date of when those actions take place. Here's the structure of the database:

Column Name	Data Type	Primary Key?	Autoinc?	Allow Null?	Unique?
id	INTEGER	YES	YES	NO	YES
time	DATETIME	NO	NO	NO	NO
event	TEXT	NO	NO	NO	NO

Does this look familiar? Yes, it's very similar to the structure of the table we created for the Tweeting Bird Feeder project. The general principles are the same: namely, we need to capture an event and record when it occurred in a structured format. This time the event is when a package arrives.

Create this table by submitting the following SQL statement to the sqlite command line:

```
[~]$ sqlite3 packagedelivery.sqlite
SQLite version 3.7.6
Enter ".help" for instructions
Enter SQL statements terminated with a ";"
sqlite> CREATE TABLE "deliverystatus" ("id" INTEGER PRIMARY KEY NOT NULL UNIQUE,
"time" DATETIME NOT NULL,"event" TEXT NOT NULL);
```

We still need a table called tracking to hold assigned tracking numbers, a description of the package contents, and the package's delivery status and the date of delivery as confirmed by the courier's own records. The structure of this table should be as follows:

Column Name	Data Type	Primary Key?	Autoinc?	Allow Null?	Unique?
id	INTEGER	YES	YES	NO	YES
tracking_number	TEXT	NO	NO	NO	NO
description	TEXT	NO	NO	NO	NO
delivery_status	BOOLEAN	NO	NO	NO	NO
delivery_date	DATETIME	NO	NO	NO	NO

Run the following SQL statement in the sqlite3 command-line tool to create this second table structure in the packagedelivery database:

```
[~]$ sqlite3 packagedelivery.sqlite
SQLite version 3.7.6
Enter ".help" for instructions
Enter SQL statements terminated with a ";"
sqlite> CREATE TABLE "tracking" ("id" INTEGER PRIMARY KEY NOT NULL UNIQUE,
"tracking_number" TEXT NOT NULL, "description" TEXT NOT NULL,
"delivery_status" BOOL NOT NULL, "delivery_date" DATETIME);
```

Now that our database tables have been created, we can proceed with the next step of obtaining the Python package dependencies we will use in the delivery processor script.

6.9 Installing the Package Dependencies

To make it easier to track packages being handled by FedEx or UPS, we are going to use a Python package called packagetrack. This wrapper helps parse the XML-formatted tracking data provided by the courier's web services, making it much easier to handle the data. While it would have been possible to use a Python screen-scraping library like Beautiful Soup, such solutions can be brittle. Not to mention, couriers the size of FedEx and UPS offer comprehensive APIs to their web services partially to discourage screen scrapers from data harvesting their sites. As such, before installing the packagetrack library, you will need to use an existing UPS and FedEx customer account to sign up for each company's service APIs. If you do not already have a customer account number and login, you will need to visit each company's website and create a new account. The account creation process requires a valid credit card number (used to bill for parcel shipping charges).

With a valid username, password, and account number in hand, visit each company's respective developer portals to sign up for a production web service API key (for FedEx) or license number (for UPS). FedEx will also generate additional security credentials (key password and meter number) when you request the production key. You will need these values to call the respective courier's web service APIs.

Next, we will install the latest packagetrack package. However, instead of retrieving it via the simple easy_install Python package retrieval and installation utility, I suggest using git to clone a fork of packagetrack maintained by Michael Stella.[3] In addition to Michael's packagetrack, download his fork of its dependency, python-fedex, which fixes a parsing issue with the FedEx XML payload.[4] The python-fedex package also relies on one more Python library dependency, called the suds library. This is a Simple Object Access Protocol (SOAP) library implementation for Python that python-fedex uses to parse the SOAP XML-wrapped payload received by the FedEx web service. Use the sudo easy_install suds Python command to automatically download and install the suds package.

---

3. https://github.com/alertedsnake/packagetrack
4. https://github.com/alertedsnake/python-fedex

Next, install both the python-fedex and packagetrack packages via the sudo python setup.py install command in the terminal window. Ensure that the packages were successfully installed by launching the Python interpreter and typing python in the terminal window. At the >>> prompt, type import packagetrack and hit return. If no error messages appeared, you installed the packages correctly.

All the other packages we will call upon in the delivery detector script are included with the standard Python 2.5 or higher distribution. With the packagetrack dependencies and courier's web service API key requirements satisfied, we are ready to write the delivery monitoring script.

6.10 Writing the Script

The package delivery monitoring script needs to perform several functions, from listening for and reacting to triggers from the package delivery monitoring hardware to sending an email alert about the event and everything in between. Specifically, the script needs to do the following:

1. Listen for threshold exceeded events (i.e., package delivery and removal) sent via the soft serial port communications between the XBee radios.

2. Timestamp these events generated by the force sensitive resistor in the deliverystatus table.

3. If a high value is received (i.e., a package is delivered), query tracking numbers stored in the tracking table. If a low value is received (i.e., a package is removed), send an email alert stating such and return execution of the script back to listening for delivery events.

4. If a high value is received, wait for a specified time before querying the tracking table to allow time for the courier to update the delivery records.

5. Iterate over undelivered tracking numbers and poll the appropriate courier's web service records for delivery confirmation status.

6. If the courier's web service results report a delivery confirmation, change the status of the tracking number record in the local tracking database table to 1 (i.e., boolean value for delivered).

7. Send an email via Gmail's secure SMTP gateway that contains the results of delivery activity in the body of the message. Note that you will need login access to an active Gmail account for this function to work.

8. Return to listening for additional package delivery events.

With those steps in mind, here's the complete script.

PackageDeliveryDetector/packagedeliverydetector.py

```python
① from datetime import datetime
  import packagetrack
  from packagetrack import Package
  import serial
  import smtplib
  import sqlite3
  import time
  import os
  import sys
  # Connect to the serial port
  XBeePort = serial.Serial('/dev/tty.YOUR_SERIAL_DEVICE', \
②   baudrate = 9600, timeout = 1)

③ def send_email(subject, message):
    recipient = 'YOUR_EMAIL_RECIPIENT@DOMAIN.COM'
    gmail_sender = 'YOUR_GMAIL_ACCOUNT_NAME@gmail.com'
    gmail_password = 'YOUR_GMAIL_ACCOUNT_PASSWORD'

    # Establish secure TLS connection to Gmail SMTP gateway
    gmail_smtp = smtplib.SMTP('smtp.gmail.com',587)
    gmail_smtp.ehlo()
    gmail_smtp.starttls()
    gmail_smtp.ehlo

    # Log into Gmail
    gmail_smtp.login(gmail_sender, gmail_password)

    # Format message
    mail_header = 'To:' + recipient + '\n' + 'From: ' + gmail_sender + '\n' \
        + 'Subject: ' + subject + '\n'
    message_body = message
    mail_message = mail_header + '\n ' + message_body + ' \n\n'

    # Send formatted message
    gmail_smtp.sendmail(gmail_sender, recipient, mail_message)
    print("Message sent")

    # Close connection
    gmail_smtp.close()

④ def process_message(msg):
    try:
      # Remember to use the full correct path to the
      # packagedelivery.sqlite file
      connection = sqlite3.connect("packagedelivery.sqlite")
      cursor = connection.cursor()

      # Get current date and time and format it accordingly
      timestamp = datetime.now().strftime("%Y-%m-%d %H:%M:%S")
```

```python
        sqlstatement = "INSERT INTO delivery (id, time, event) \
VALUES(NULL, \"%s\", \"%s\")" % (timestamp, msg)
        cursor.execute(sqlstatement)
        connection.commit()
        cursor.close()
    except:
        print("Problem accessing delivery table in the " \
        + "packagedelivery database")

    if (msg == "Delivery"):

        # Wait 5 minutes (300 seconds) before polling the various couriers
        time.sleep(300)

        try:
            connection = sqlite3.connect("packagedelivery.sqlite")
            cursor = connection.cursor()
            cursor.execute('SELECT * FROM tracking WHERE '\
            + 'delivery_status=0')
            results = cursor.fetchall()
            message = ""

            for x in results:
                tracking_number = str(x[1])
                description = str(x[2])
                print tracking_number

                package = Package(tracking_number)
                info = package.track()
                delivery_status = info.status
                delivery_date = str(info.delivery_date)

                if (delivery_status.lower() == 'delivered'):
                    sql_statement = 'UPDATE tracking SET \
                    delivery_status = "1", delivery_date = \
                    "' + delivery_date + \
                    '" WHERE tracking_number = "' \
                     + tracking_number + '";'
                    cursor.execute(sql_statement)
                    connection.commit()
                    message = message + description \
                    + ' item with tracking number ' \
                    + tracking_number \
                    + ' was delivered on ' \
                    + delivery_date +'\n\n'

            # Close the cursor
            cursor.close()
```

```
                  # If delivery confirmation has been made, send an email
                  if (len(message) > 0):
                    print message
                    send_email('Package Delivery Confirmation', message)
                  else:
                    send_email('Package Delivery Detected', 'A ' \
                    + 'package delivery event was detected, ' \
                    + 'but no packages with un-confirmed ' \
                    + 'delivery tracking numbers in the database ' \
                    + 'were able to be confirmed delivered by ' \
                    + 'the courier at this time.')
                except:
                  print("Problem accessing tracking table in the " \
                  + "packagedelivery database")

            else:
                send_email('Package(s) Removed', 'Package removal detected.')

⑤ if sys.platform == "win32":
        os.system("cls")
  else:
        os.system("clear")

  print("Package Delivery Detector running...\n")
  try:
    while 1:
        # listen for inbound characters from the XBee radio
        XBee_message = XBeePort.readline()

        # Depending on the type of delivery message received,
        # log and lookup accordingly
        if "Delivery" in XBee_message:
          # Get current date and time and format it accordingly
          timestamp = datetime.now().strftime("%Y-%m-%d %H:%M:%S")
          print("Delivery event detected - " + timestamp)
          process_message("Delivery")

        if "Empty" in XBee_message:
          # Get current date and time and format it accordingly
          timestamp = datetime.now().strftime("%Y-%m-%d %H:%M:%S")
          print("Parcel removal event detected - " + timestamp)
          process_message("Empty")

⑥ except KeyboardInterrupt:
    print("\nQuitting the Package Delivery Detector.\n")
    pass
```

① Begin by importing the script's dependencies on the custom packagetrack library along with the serial, smtplib, sqlite3, time, os, and sys standard Python libraries.

② Identify the serial port of the XBee radio attached to the computer's serial port. This radio will listen for incoming transmissions from the paired XBee radio attached to the Arduino connected to the force sensitive resistor. Replace the '/dev/tty.YOUR_SERIAL_DEVICE' placeholder with the actual path of the your XBee radio's attached serial port value.

③ This is the send_mail routine that is used in the process_message routine, and thus it needs to be declared first. Replace the recipient, gmail_sender, and gmail_password placeholder values with your desired recipient and Gmail account credentials.

④ The process_message routine is where most of the action happens in the script. Connect to the packagedelivery.sqlite SQLite database and log the type of event received. If a delivery message is received, the script waits for five minutes before polling the FedEx and UPS web services to give the shipper enough time to log the delivery status to the central servers. Then, the tracking table in the packagedelivery.sqlite database is queried for any undelivered tracking numbers. These numbers are submitted one at a time to the respective web service. If a delivery confirmation is returned, its positive response is logged to the database with the confirmed delivery date, as well as appended to the body of the email message to be sent via the send_email routine.

⑤ This is the main loop of the script. Begin by clearing the screen and listening for a "Delivery" or "Empty" message from the Arduino-attached XBee radio and invoke the process_message routine.

⑥ Gracefully exit the script if a Ctrl-C keypress is detected.

Save the script as packagedelivery.py and execute it with the python packagedelivery.py command. If any errors arise, check syntax and code-line indentations, since Python is very strict about line formatting. If the script starts up without any complaints, you're ready to test it out.

6.11 Testing the Delivery Processor

With the Python script written and either (or both) FedEx and UPS customer accounts and web developer API keys registered, we can now proceed with running the sketch and listener script through a functional test. Load a valid, recent FedEx or UPS tracking number into the trackingstatus table in the packagedelivery database. To do this, you can use the same SQLite Manager plug-in for Firefox that was used to create the database tables. Simply select SQLite Manager from Firefox's Tools menu, then open the packagedelivery.sqlite file. Click the trackingstatus table listed in the left column area, followed by the Browse & Search tab. Lastly, click the Edit button to add/modify the tracking number record(s).

Conversely, if you prefer the faster (though less visually stimulating) sqlite3 command-line interface, add your own tracking number(s) via the following SQL statement (remembering, of course, to replace the YOURTRACKINGNUM placeholder with a valid FedEx or UPS tracking number):

```
sqlite> INSERT INTO tracking("tracking_number","description","delivery_status")\
VALUES ("YOURTRACKINGNUM", "My Package Being Tracked","0");
```

Quickly check that the record was indeed correctly added to the tracking table with a simple select statement:

```
sqlite> select * from tracking;
1|YOURTRACKINGNUM|My Package Being Tracked|0|
```

Add any other valid FedEx or UPS tracking numbers as well to test the iterative lookup functionality that was coded into the Python script. For the purposes of this test, tracking numbers don't have to only be those for packages in transit. In fact, it's best to have a mix of both in-transit and delivered packages to verify that the script correctly updated only those packages with tracking numbers that have a confirmed delivery status.

The moment of truth has arrived. Power up the Arduino/XBee package delivery hardware. Make sure the receiving XBee is plugged into your computer via the FTDI cable and execute the python packagedelivery.py command. Press firmly on the force sensitive resistor and wait for the script to process the tracking number queries. If everything worked successfully, you should have received an email from the Gmail account you used for the SMTP gateway that listed all confirmed package deliveries. You can also execute another select * from tracking; query from the sqlite3 command line to verify that the boolean delivered field has been changed from 0 (false) to 1 (true), and that the appropriate time stamp (indicating when the package was actually delivered) was recorded in the deliver_time field.

If the script failed or if the values were not properly stored, reset the records in the tracking table and use a variety of debugging approaches with the script (the easiest being the venerable print() function) to determine where problems are arising in the script's execution.

Once the tests have successfully and repeatedly passed, we're ready to install the hardware configuration outdoors in a convenient location close to a power outlet.

6.12 Setting It Up

First, identify an appropriate spot to place the pressure plate. Most parcel delivery companies drop off packages at a resident's front door, just off to the left or right side so as not to block the entrance. To help guide to placement of the packages, you can leave a note or sign for the courier to specifically drop boxes on the designated rectangular pressure plate we built for the project.

Better yet, use or build a container with a lid and place the sensor-embedded plate on the bottom of the container. You can purchase inexpensive, sturdy, water-resistant containers that also act as large seats when the top lid is down. With a little extra work, you can place the Arduino and XBee radio in a hard-shelled, waterproof enclosure and carefully mount the assembly to the inside of the container.

Tuck it far enough to the side so that it does not obstruct any packages that might forcibly land on the assembly and possibly damage the electronics. Post a note asking delivery personnel to place packages in the container. Depending on how frequently you receive parcels, this new behavior may be adopted quickly by those who manage the routes in your area.

Due to the proximity of the detector to the front entrance of the home, an outdoor power outlet should be easy to locate and use. If the placement of your assembled detector happens to be in direct sunlight for most of the day, you can even try powering the electronics via a solar cell battery like the one recommended for the Tweeting Bird Feeder project.

With everything set up and powered, test out the detector for yourself. Try boxes of various shapes, sizes, and weights to see how the configuration reacts to each. You may need to reposition the force sensitive resistor to achieve a consistent trigger. Adding a second and even a third force sensitive resistor will also greatly improve detection, especially for smaller packages that may not hit the center sensor plate.

Your Package Delivery Detector is now complete and ready to process deliveries. Place orders with your favorite online retailers to see how much more convenient and reassuring it is to know that an anticipated package is awaiting your retrieval when you get home.

6.13 Next Steps

It is easy to extend the detector beyond package delivery notification. Here are a few ideas that can be used to further develop the concept.

- The current design is limited to one package delivery before it needs to be reset. Enhance the sketch and Python script to account for multiple package deliveries from multiple couriers. For instance, if a courier delivers a package that triggers a threshold event, set a new threshold value such that another delivery can be detected and verified before you retrieve the first package.

- When the force sensitive resistor's threshold is exceeded, capture a photo of the delivery in progress and send it as an attachment with the delivery email confirmation.

- Enhance the database portion of the Python script to store the results of the package delivery query and write a web front end in Django to use as a package delivery history tool.

- Place the force sensitive resistor under your front doormat and be alerted when a visitor comes calling before the doorbell rings. To remotely unlock the door for trusted individuals, combine with a webcam and an electric lock from Chapter 9, *Android Door Lock*, on page 141.

- If you frequently receive packages, employ message notification alternatives beyond email and Twitter. Write a package delivery notification service for Android or an iMessage-enabled delivery app for iOS that will natively alert you about package arrivals.

- Shrink the delivery detector assembly with the same Arduino Nano/XBee configuration used in Chapter 5, *Tweeting Bird Feeder*, on page 57. Swap out the force sensitive resistor with a PIR to notify you when a hand reaches into your mailbox.

- Combine the detector with the guard dog from Chapter 4, *Electric Guard Dog*, on page 43. Use the force sensitive resistor to notify the guard dog to power up and look for movement. Combine with a laser and range finder to accurately "paint the target" to greet visitors with a high-tech welcome.

CHAPTER 7

Web-Enabled Light Switch

Imagine coming home after a long day at work and being able to power on lights, televisions, and appliances simultaneously from a native control application running on your mobile phone. Any electrical device with a standard power plug can be a part of this vision.

We're going to make that vision a reality with the help of a networked computer, a Ruby on Rails web application, a native Android phone application, and an older home automation technology known as X10. We will build a native remote light switch Android application that can turn lights on and off with the touch of an onscreen toggle switch (Figure 21, *Easily control your home's lighting and electrical appliances*, on page 106). When we're finished, we'll have the ability not only to control appliances nearby but also anywhere there is an Internet connection, should we decide to make the Rails server publicly accessible.

7.1 What You Need

X10 is a company that has been selling its proprietary electrical switches for many years, and the technology has changed little since its introduction over thirty years ago. Yet regardless of its age, X10 power switches are still a primary home automation technology, mostly because they are inexpensive and, when coupled with a computer, can send and schedule power on/off events.

Instead of relying on X10's rigid and proprietary Windows-based application to control X10 devices, we are going to use a freely available open source utility called Heyu. Created and maintained by Daniel Suthers and Charles Sullivan, Heyu provides a command-line interface to monitor and send a variety of X10 commands to the CM11A. These instructions will then be relayed to the specified X10 switches.

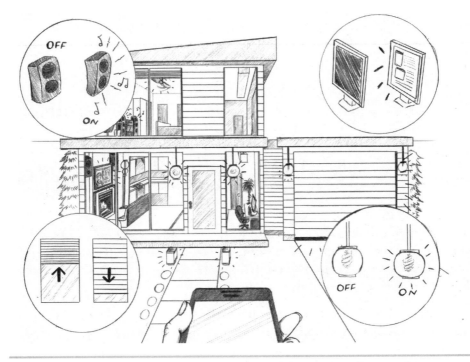

Figure 21—Easily control your home's lighting and electrical appliances ...from your own custom smartphone application.

For this project, it is best to stick with the Linux or Mac operating system, since they can easily compile the source code without modification. Unfortunately, there is no native port of Heyu available for Windows, and none is planned anytime in the near future. If you are using Windows, consider running a Linux distribution in a virtual machine using a program like VirtualBox.[1]

You will need the following parts (refer to the photo in Figure 22, *Web-Enabled Light Switch parts*, on page 107):

1. X10 CM11A computer interface[2]—note that unlike the serial port-based CM11A, the newer X10 CM15A model connects to a computer via USB and will not work with Heyu software. See the Heyu FAQ for more details.[3]
2. X10 PLW01 standard toggle wall switch
3. Serial to USB interface cable

---

1. https://www.virtualbox.org/
2. http://www.x10.com
3. http://www.heyu.org/heyu_faq.html

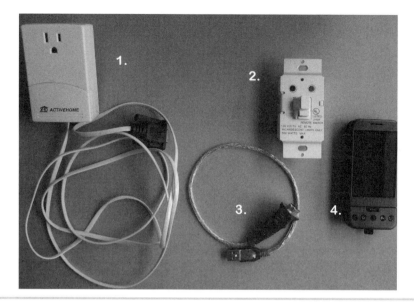

Figure 22—Web-Enabled Light Switch parts

4. An Android OS phone or tablet device (used to run the Web-Enabled Light Switch client application)
5. A computer (not shown), preferably Linux or Mac-based, with Ruby 1.8.7 or higher installed

Additionally, you will need the following software:

- Heyu 2.9.3 or higher[5]
- Ruby on Rails 3.0 or higher[6]
- The Eclipse IDE[7]
- The Android SDK 1.5 or higher[8]
- The Android Development Tools (ADK) Plugin for Eclipse[9]

At the heart of any computer-assisted X10 setup is the control module. The module provides the interface between the transmission of instructions to X10 devices as well as for the notifications of triggers (ex: motion detection) from X10 appliances equipped with such capabilities. Several of these interfaces exist, such as the X10 Firecracker (known by its serial number as the

---

5. http://heyu.org
6. http://www.rubyonrails.com
7. http://eclipse.org
8. http://developer.android.com/sdk
9. http://developer.android.com/sdk/eclipse-adt.html

> **How Does X10 Work?**
>
> The basic premise behind X10 is sending unique pulse codes over existing electrical wiring to devices capable of acting on those codes. Each device is manually set to its own unique house code and device code, H8 for example. To activate an X10 power switch, this unique identifier is transmitted from a control interface plugged into a power outlet. Using the home's existing electrical wiring, the house code leaves the control interface as a series of pulses that traverses over the wiring. The X10 device that is set to the target code recognizes that it is the recipient of the code that follows.
>
> The function code can consist of a simple on or off message that in turn triggers the relay in the receiving X10 module. This in turn switches on or off the power going to the lamp or appliance plugged into the module. Besides the basic on and off codes, other instructions can be sent as well, such as setting a light dimmer to 25 percent or even turning all X10 devices on or off simultaneously. For a more comprehensive explanation and a list of these codes, visit the Heyu website.[a]
>
> ---
> a. http://www.heyu.org/docs/protocol.txt

CM17A) or the original X10 computer interface, the CM11A. Most of the open source X10 automation software available today supports both of these and other interfaces, but I find the CM11A to be the most prevalent. Hence, I recommend using the CM11A for this project.

With the required hardware and software in hand, let's take a look at how we are going to combine all this technology to make it turn a light on and off from an Android smartphone application.

7.2 Building the Solution

In order for X10-managed lights and appliances to be remotely controlled, we are going to assemble a variety of separate technologies and use them in a unified way. We will do the following:

1. Test the X10 computer interface and modules with the Heyu application.

2. Create a Ruby on Rails application that provides a web-based front end to a subset of Heyu commands.

3. Create an Android mobile application that will communicate with the Rails application, turning the light on and off via a native onscreen Android toggle switch control.

We will start by hooking up the X10 hardware and verifying that it can be controlled via the Heyu application.

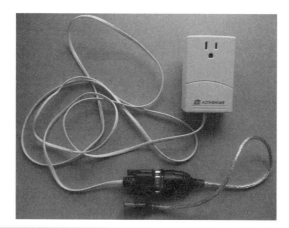

Figure 23—The X10 CM11A interface controls the Web-Enabled Light Switch.

7.3 Hooking It Up

Plug the X10 CM11A into an outlet near your computer so that its interface cable is within reach of your computer. Because the CM11A uses a 9-pin serial connection, you will need a USB-to-serial adapter and the appropriate driver, similar to the illustration shown in Figure 23, *The X10 CM11A interface controls the Web-Enabled Light Switch*, on page 109. If you use a Mac running OS version 10.6 or higher, you can download the PL-2303 driver from the Prolific website.[10] If the Prolific website is unavailable, you can obtain unofficial drivers from Bryan Berg's fork of the drivers on Github.[11] Computers running the latest Linux distributions should have no trouble identifying and connecting to the PL-2303 interface.

Next, plug the USB-to-serial adapter into a USB port on your computer and attach it to a powered CM11A interface. You'll need to interrogate the device for the serial port that the operating system assigned to the interface. You can do this via locating the appropriate tty device file in the /dev directory by issuing a ls /dev/tty* command in the terminal window. Easier still, load up the Arduino IDE and select the Tools→Serial Port menu option. In my case, the device name of the CM11A is /dev/tty.usbserial, as shown in Figure 24, *The Arduino Tools menu displaying the USB serial adapter name*, on page 110. Note the name of the new device path since you will need to refer to it in the Heyu configuration file.

---

10. http://www.prolific.com.tw/eng/downloads.asp?ID=31I.
11. https://github.com/failberg/osx-pl2303.

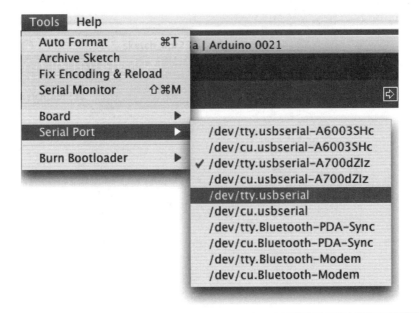

Figure 24—The Arduino Tools menu displaying the USB serial adapter name

Now that the CM11A is connected to and recognized by your computer, download the Heyu source code from the Heyu.org website, uncompress the tarball via the tar -zxvf heyu-2.9.3.tar.gz command. Then, perform a ./Configure;make;make install cycle to install the compiled application. If you use a Mac, you need to have the Mac developer tools installed before proceeding.[12] If you are using a Linux computer, make sure you have the necessary gcc compile and make tools installed.

For example, if you are using a Debian-based Linux distribution like Ubuntu, issue sudo apt-get install build-essential from the terminal window to download and install the compiler and linker tools. Then, execute the usual ./Configure, make, and sudo make install to compile the source and install the heyu executable and dependencies on your computer.

Along with the heyu executable, a x10.conf configuration file is installed in the /etc/heyu directory. Open this file for read-write editing (ex: sudo vi /etc/heyu/x10.conf).

There are a number of options that can be set in the x10.conf file, but the one we're most concerned with now is the serial port path to the CM11A that you identified earlier.

---

12. http://developer.apple.com/technologies/tools/

```
# Serial port to which the CM11a is connected.  Default is /dev/ttyS0.
TTY      /dev/tty.usbserial
```

Use your favorite text editor to enter the serial port value that matches the location of your CM11A device. Save the file and test the settings by first launching the Heyu state engine via a terminal window with the command:

```
> heyu engine
```

If no errors are reported, you're in good shape since the engine daemon found the device and is running successfully in the background. You can also try entering heyu info for more details about the Heyu configuration. Now, in the same terminal enter this:

```
> heyu monitor
```

This will monitor the interaction of the CM11A with other X10 devices. Assuming you have set the housecode of the PLW01 wall switch to H3, enter this command:

```
> heyu on h3
```

That should turn on the switch and complete the circuit for whatever electrical device (such as a ceiling lamp) it is routing electricity to. You should also see the terminal window running the Heyu monitor process report the transmittal of the command:

```
07/25 12:45:34  sndc addr unit         3 : hu H3  (_no_alias_)
07/25 12:45:34  sndc func             On : hc H
```

Turn off the switch by issuing an off command to the H3 device:

```
> heyu off h3
07/25 12:50:17  sndc addr unit         3 : hu H3  (_no_alias_)
07/25 12:50:18  sndc func            Off : hc H
```

If these commands fail to turn the switch on and off, try another X10 module, like an AM486 Appliance module. If that also fails, try bringing the wall switch closer to the X10 computer interface, preferably on the same room wiring.

The majority of issues I have encountered with X10 projects are often directly associated with the reliability of the fire-and-forget protocol of X10 itself. If you suspect the problem may be the X10 hardware, try swapping the X10 device in question with replacement hardware. You can also have an electrician check for line noise or electrical wiring issues that may be hindering the transmission of X10 pulses from the CM11A interface to the accompanying X10 modules.

> **X10 Problems**
>
> While X10 is the most prevalent (and most advertised) low-cost home automation solution available today, it does have a number of constraints. Besides the problems with its fire-and-forget protocol (i.e., X10 sends out messages but has no way to verify whether or not the device acknowledged and acted upon the request), the other big problem with X10 is its use of home electrical wiring to propagate its signals.
>
> Home wiring often is both noisy and degrades over time. Such wiring connectivity can be exacerbated by X10 modules plugged into surge-protecting power strips and other line-conditioning end points that can filter out the intentional fluctuations that X10 commands dump into the electrical stream. Depending on how far the X10 signal needs to travel, additional X10 modules may be required to ensure that X10 commands reach their intended destination. Yet even with these annoyances, X10 remains one of the most cost-effective and easiest home automation solutions to implement. Even though a number of competing alternatives to X10 have been introduced over the thirty-plus years that X10 has been commercially available, none have yet matched X10's low cost and ease of implementation.

Once the computer and CM11A are talking to one another via the heyu command-line interface, we can leverage this functionality by encapsulating it into a web application. This way, we can easily access and control X10 end points from a web browser and ultimately from an Android application.

7.4 Writing the Code for the Web Client

For the Web-enabled light switch, we will create a simple Ruby on Rails project to manage the user interface interaction first via a web browser. We won't spend a lot of time on the user interface, though, since that will ultimately be the job of the custom Android application we will create after the web interface is functionally tested.

Rails runs optimally on Mac or Linux computers, and it is already installed by default on Mac OS X 10.6. However, it is not the latest version. Because this project requires Rails 3.0 or higher, the instructions are not applicable to older versions of the framework. Follow the instructions on the Ruby on Rails website to get the latest Rails release running on your computer.

With the Rails web framework installed, create a new directory and switch to that directory before creating the new Rails project:

```
> mkdir ~/projects/ruby/rails/homeprojects/
> cd ~/projects/ruby/rails/homeprojects
> rails new x10switch
```

```
      create
      create  README
      create  Rakefile
      create  config.ru
      create  .gitignore
      create  Gemfile
      create  app
      create  app/controllers/application_controller.rb
      create  app/helpers/application_helper.rb
      create  app/mailers
      create  app/models
      ...
      create  vendor/plugins
      create  vendor/plugins/.gitkeep
```

Next, change into the new x10switch directory and create a new controller called command with an action called cmd() to manage the interaction between the web interface and the Heyu terminal application.

```
> cd x10switch
   > rails generate controller Command cmd

      create   app/controllers/command_controller.rb
       route   get "command/cmd"
      invoke   erb
      create    app/views/command
      create    app/views/command/cmd.html.erb
      invoke   test_unit
      create    test/functional/command_controller_test.rb
      invoke   helper
      create    app/helpers/command_helper.rb
      invoke   test_unit
      create     test/unit/helpers/command_helper_test.rb
```

Then, locate the app/controllers/command_controller.rb file and check for the on and off parameters and execute the appropriate action:

```ruby
class CommandController < ApplicationController
  def cmd
    @result = params[:cmd]

    if @result == "on"
      %x[/usr/local/bin/heyu on h3]
    end

    if @result == "off"
      %x[/usr/local/bin/heyu off h3]
    end
  end
end
```

The %x is a Ruby construct to execute an application with command-line arguments. Hence, %x[/usr/local/bin/heyu on h3] tells Heyu to send an on command code to the H3 house code X10 switch. Likewise, the %x[/usr/local/bin/heyu off h3] tells that same switch to turn off.

Next, edit the app/views/command/cmd.html.erb document and replace its placeholder contents with the following single line of embedded Ruby code to display the results of the On and Off request:

```
The light should now be <%= @result %>.
```

While we could go much further with this Rails application, dressing it up with a nice user-friendly interface accessed from the public/index.html file as well as providing more verbose output of the result of the action, I will leave that exercise for the aspiring reader. Since we will ultimately be controlling the switch from a native mobile client application, there's little incentive to invest time in whipping up a sparkly web UI when it will hardly ever be seen.

Finally, edit the config/routes.rb file and replace the get "command/cmd" with the following:

```
match "/command/:cmd", :to => 'command#cmd'
```

This instructs the Rails application on how to route incoming command requests to execute the on/off actions. Save your work and get ready to rumble!

If you're setting up a newer version of Rails (such as Rails 3.1) on a Linux system, you may also need to install a few package dependencies (or gems as they're known in Ruby parlance) in order for Rails to run. Just edit the Gemfile file that was generated in the x10switch directory and add the following:

```
gem 'execjs'
gem 'therubyracer'
```

Save the changes and then run this command:

```
> bundle install
```

This will download and install the extra files used by the Rails 3.1 JavaScript processing engine. With these two gems successfully installed, you're ready to run and test out the X10switch Rails application.

7.5 Testing Out the Web Client

With the X10 computer interface working and plugged into the serial port of the computer, fire up a development server of the Rails 3 code via this:

```
> cd ~/projects/ruby/rails/homprojects/x10switch
> rails s

=> Booting WEBrick
=> Rails 3.0.5 application starting in development on http://0.0.0.0:3000
=> Call with -d to detach
=> Ctrl-C to shutdown server
[2011-03-18 16:49:31] INFO  WEBrick 1.3.1
[2011-03-18 16:49:31] INFO  ruby 1.8.7 (2009-06-12) [universal-darwin10.0]
[2011-03-18 16:49:31] INFO  WEBrick::HTTPServer#start: pid=10313 port=3000
```

Open a web browser on the local machine and enter the following:

http://localhost:3000/command/on

If everything is coded correctly, you should see The light should now be on. in the browser window, as shown in Figure 25, *The browser should indicate the proper status of the light*, on page 116.

More importantly, Heyu should have executed the on command for the X10 device coded with the H3 house code. In other words, the light should have turned on. Turn the light off by submitting the off command:

http://localhost:3000/command/off

If the light turned off, congratulations! You have wired up and programmed everything correctly. When you're ready to expand the Rails application to handle even more commands, just add more if @result == statements to the CommandController class containing the command you want Heyu to transmit. These commands could range from dimming lights to 30 percent, turning an appliance on for a specified duration, or managing a combination of power on/off events.

If you're interested in learning more about programming web applications using the Ruby on Rails framework, check out *Programming Ruby: The Pragmatic Programmer's Guide* [TFH09].

Now that the web application server is working, it's time to build a mobile client.

7.6 Writing the Code for the Android Client

You might be wondering why you should go through the trouble of building a native Android client when the web application we wrote can be accessed by the Android mobile web browser. Well, if all you wanted to do was toggle light switches on and off, then I would say you don't need a native client. The web interface works just fine and can be further enhanced using AJAX and

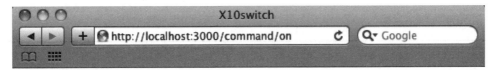

The light should now be on.

Figure 25—The browser should indicate the proper status of the light.

slick HTML5/CSS3 user interface effects. But if you want to give a little more intelligence to the app, such as activating power switches based on your proximity to them or running an Android service that monitors inbound X10 events like motion detection and then sounds an alert on your phone to bring such events to your attention, a dynamic web page just won't do.

If you haven't already done so, download, install, and configure the Eclipse IDE, the latest Android SDK, and the ADK plug-in for Eclipse. Visit the Android SDK website for details on how to do so.[13]

You will also need to create an Android Virtual Device (AVD) so that you can use it to test the client application in an Android emulator before sending the program to your Android device.[14] I suggest creating an AVD that targets Android 1.5 (API Level 3) so that it emulates the largest number of Android phones available.

Launch the Eclipse environment and select File→New→Android Project. Depending on the version of Eclipse you are running, this option might also be found on the File menu via New->Other->Android->Android Project. Call the project LightSwitch and select Build Target as Android 1.5. You can choose a higher Android version depending on what level of Android device you want to deploy the application to, but since the LightSwitch program will be sweet and simple, Android 1.5 should be adequate for this sample application.

In the Properties area, fill in the Application name as Light Switch and the Package name as com.mysampleapp.lightswitch, and check the Create Activity checkbox and enter LightSwitch. You can specify the Min SDK Version if you wish, but since we're developing for one of the more popular lowest-common-denominator versions of Android, we'll leave it blank for now. Before you continue, check to see if your New Android Project dialog box looks like the one shown in Figure 26, *Creating a new Android Project dialog box with completed parameters*, on page 117.

---

13. http://developer.android.com/sdk
14. http://developer.android.com/guide/developing/devices/managing-avds.html

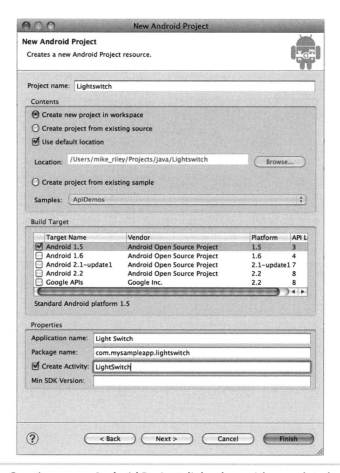

Figure 26—Creating a new Android Project dialog box with completed parameters

Android developers with good testing practices would then click the Next button in the New Android Project dialog box to set up a Test Project resource. However, in the interest of space and time, we'll go ahead and click the Finish button.

Once the Android Development Tools Eclipse plug-in generates the skeleton Light Switch application code, double-click the main.xml in the res/layout folder to open it into Android's simple form editor. Drag a ToggleButton control from the Form Widgets palette onto the main.xml graphical layout. Don't worry about perfectly aligning the control in the right spot for now. For this exercise, we're more interested in function over form.

Because this application won't require anything beyond the basic features found in the earlier Android operating system releases, change the Android

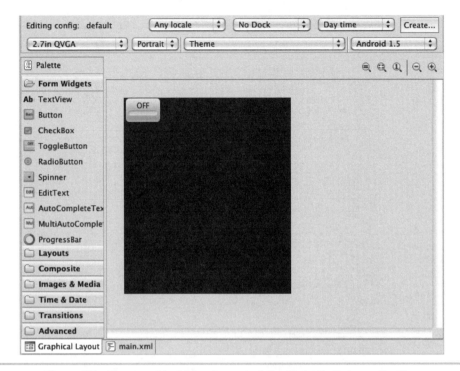

Figure 27—The graphical form layout of the Light Switch application

version in the upper right corner drop-down box of the form editor to Android 1.5. Also, feel free to delete the default Hello world TextView element from the layout. When done, the layout should look similar to the screen shown in Figure 27, *The graphical form layout of the Light Switch application*, on page 118. Save the main.xml file.

Expand the src→com.mysampleapp.lightswitch tree and double-click the LightSwitch.java file. Because we will be using the ToggleSwitch widget, the first thing we need to import is the android.widget.ToggleButton class.

Next, add the java.net.URL and java.io.InputStream libraries, since we'll be creating URL objects to pass to Java InputStream object. The import statement section of the LightSwitch.java file should now look like this:

```
package com.mysampleapp.lightswitch;
import android.app.Activity;
import android.os.Bundle;
import android.widget.ToggleButton;
import android.view.View;
import java.net.URL;
import java.io.InputStream;
```

Now we have to make the LightSwitch aware of the ToggleSwitch by finding it by ID in the LightSwitch class's OnCreate event and adding an event listener to monitor when the switch is toggled on and off:

```
public class LightSwitch extends Activity {
  /** Called when the activity is first created. */
    @Override
    public void onCreate(Bundle savedInstanceState) {
      super.onCreate(savedInstanceState);
      setContentView(R.layout.main);
      final String my_server_ip_address_and_port_number =
        "192.168.1.100:3344";
      final ToggleButton toggleButton =
          (ToggleButton) findViewById(R.id.toggleButton1);
      toggleButton.setOnClickListener(new View.OnClickListener()
      {
        public void onClick(View v) {
          if (toggleButton.isChecked()) {
            try {
                final InputStream is = new URL("http://"+
my_server_ip_address_and_port_number +"/command/on").openStream();
            }
            catch (Exception e) {
            }
          } else {
            try {
                final InputStream is = new URL("http://"+
my_server_ip_address_and_port_number +"/command/off").openStream();
            }
            catch (Exception e) {
            }
          }
        }
      });
    }
}
```

Be sure to set the my_server_ip_address_and_port_number string in the example above to the IP address and port that you plan to use to run the Rails application server we wrote in Section 7.4, *Writing the Code for the Web Client*, on page 112. And that's it! Go ahead and run the application in the Android emulator to make sure it compiles and shows up on the screen correctly.

7.7 Testing Out the Android Client

Time to test out the application on a real X10 light switch. Assuming the Rails-based X10 web application is working as expected, start up your Rails development server on the same network/subnet as the emulator in Figure 28, *Running the Light Switch application*, on page 121.

Use the same port number as the one we assigned in the my_server_ip_address_and_port_number string from our Android application. For example, in the case of 192.168.1.100:3344, the IP address is 192.168.1.100 and the port number is 3344. Pass this as a command-line parameter when launching the Rails server instance, like this:

```
> rails s -p3344
```

With the rails development server now running on port 3344 and waiting for inbound requests on the same local area network as your Android emulator or device, click the On/Off toggle button.

Um, nothing happened. Why?

There is one more important setting we have to make in the Light Switch application configuration. We have to respect the Android application security model and tell the Android OS that we want to allow our application to Use the Internet so that we can have our outbound HTTP requests reach the outside world. To do so, double-click the AndroidManifest.xml file and add the following line just above the closing manifest tag, like this:

```
<uses-permission android:name="android.permission.INTERNET">
</uses-permission>
```

The entire AndroidManifest.xml file should now look like this:

```
<?xml version="1.0" encoding="utf-8"?>
<manifest xmlns:android="http://schemas.android.com/apk/res/android"
   package="com.mysampleapp.lightswitch"
   android:versionCode="1"
   android:versionName="1.0">
    <application android:icon="@drawable/icon"
                 android:label="@string/app_name">
      <activity android:name=".LightSwitch"
                android:label="@string/app_name">
        <intent-filter>
          <action android:name="android.intent.action.MAIN" />
          <category android:name="android.intent.category.LAUNCHER" />
        </intent-filter>
      </activity>
    </application>
    <uses-permission android:name="android.permission.INTERNET">
    </uses-permission>
</manifest>
```

Figure 28—Running the Light Switch application

Recompile and run the Light Switch application with the new permission setting and click the toggle button. If everything worked as planned, you should see the Rails server report something similar to the following successfully received request:

```
Started GET "/command/on" for 192.168.1.101 at Sat Mar 21 19:48:10 -0500 2011
  Processing by CommandController#cmd as HTML
  Parameters: {"cmd"=>"on"}
Rendered command/cmd.html.erb within layouts/application (11.7ms)
Completed 200 OK in 53ms (Views: 34.7ms | ActiveRecord: 0.0ms)
```

You should also see the light turn on! Click the toggle button again. It should generate a similar report for the off command:

```
Started GET "/command/off" for 192.168.1.101 at Sat Mar 26 19:52:30 -0500 2011
  Processing by CommandController#cmd as HTML
  Parameters: {"cmd"=>"off"}
Rendered command/cmd.html.erb within layouts/application (13.2ms)
Completed 200 OK in 1623ms (Views: 40.0ms | ActiveRecord: 0.0ms)
```

Consequently, the light should now switch off.

On rare occasions, one other issue you may encounter when you attempt to install the Light Switch application on your Android phone is an expired debug key. Android's security model requires a signed key to execute code on an Android device. The signed key should have been automatically generated and configured when you installed the Android SDK, but in the event that an expiration message occurs, follow the signing procedure in the Android SDK documentation to generate a new key.[15]

For more details on installing Android programs onto an Android device from the Eclipse environment, review the Android SDK documentation on running Android applications on an emulator and on a device.[16]

7.8 Next Steps

Congratulations! Now that you can control a lamp or appliance via a graphical toggle switch on your Android phone, a whole new world of home automation possibilities awaits.

You could continue to enhance the native mobile client by controlling multiple X10 switches in a facile and elegant manner based on the time of day and the GPS coordinates of your mobile device (ex: turn on the porch light after dark when you're within a five-meter radius of the front door). While we won't be building this location-based application in this book, you have the basic building blocks already in your possession. If you're interested in going further, read Ed Burnette's *Hello, Android* [Bur10] for some really helpful tutorials, then post your ideas and creations on the Programming Your Home website!

Several other improvements can be made to the configuration to make the system more robust and user friendly. These include the following:

- Improve application error trapping and reporting. X10 methods are "fire and forget" events that do not inherently return success or failure. As such, there are plenty of enhancements that can be made to the web service to troubleshoot nearly everything else up to the point of X10 command transmissions. Before sending a message, trap for and report on X10 computer interface connection errors. Knowing that the X10 interface is down is far more helpful than seeing no action from the switch without any explanation of why.

---

15. http://developer.android.com/guide/publishing/app-signing.html
16. http://developer.android.com/guide/developing/building/building-eclipse.html

- Invest in several more X10 modules, ranging from two-way (LM14A) and the Socket Rocket lamp (LM15A) to wall socket (SR227) and heavy-duty appliance (HD243) modules. Send Heyu transmission events simultaneously to multiple X10 devices. For example, using a single method call, turn on lights in the kitchen; power up the toaster, coffee maker, and ceiling fan; and report back to the mobile client when coffee will be brewed and a toasted bagel will be ready.

- If you prefer a more lightweight Ruby-based web framework, consider replacing the Ruby on Rails server application with one using Sinatra.[17] While it hasn't yet matched the popularity of Rails, Sinatra is nevertheless a pretty nifty minimalist Ruby-based framework that is worthy of a closer look.

- Dress up the mobile user interfaces with a more elegant, multifunctional front end that can be used for multiple web-enabled switches, appliances, garage door openers, and more.

- Extend functionality to other projects, such as an Arduino-based TV remote, or link together interface controls to other projects in this book, such as Chapter 8, *Curtain Automation*, on page 125, and Chapter 9, *Android Door Lock*, on page 141.

---

17. http://www.sinatrarb.com/

CHAPTER 8

Curtain Automation

One of the more frequent effects in science fiction movies about home life is the autonomous opening and closing of curtains and window shades. Well, the future is here and it's about to get more evenly distributed. In this project, we will construct a system that will open and close curtains based on light and temperature. When the heat goes up, the curtains close. Likewise, when the sun comes up, the curtains open (Figure 29, *Automate curtains and shades*, on page 126).

To bring motion to this solution, our primary hardware component will be a stepper motor, a continuous rotational engine that will be driven by an Arduino to spin a certain number of revolutions clockwise and counterclockwise. When the shaft of the stepper motor is connected to a curtain string and pulley system, the motor will open and close the curtains accordingly.

Let's take a look at the other supplies we will need to build this project.

8.1 What You Need

The parts required for this project are fairly straightforward. Sensors for light and temperature, a stepper motor, and an Arduino board are the primary components. Then it's just a matter of mounting and powering the assembly. Refer to the photo in Figure 30, *Curtain Automation parts*, on page 127. Specifically, we will need the following:

1. Four 2-inch angle brackets to mount the stepper motor in place
2. A 12V bipolar stepper motor[1]
3. Arduino motor shield[2]

---

1. https://www.adafruit.com/products/324
2. http://www.adafruit.com/products/81

Figure 29—**Automate curtains and shades** ...depending on light and temperature.

4. A 12V power supply[3]
5. Double-sided foam tape to dampen the vibration of the mounted stepper motor
6. A grooved rubber pulley wheel to grip and move the curtain drawstring
7. Wire to connect the sensors and stepper motor to the motor shield
8. A TMP36 analog temperature sensor (for a close-up image of a photocell and a temperature sensor, see Figure 31, *Curtain Automation sensors*, on page 128)[4]
9. A 10k ohm resistor (usually coded with brown, black, orange, and gold bands)
10. A photoresistor (the same type we used in Chapter 5, *Tweeting Bird Feeder*, on page 57)
11. An Arduino Uno
12. A small breadboard to mount the photocell and temperature sensors
13. A standard A-B USB cable (not pictured) to connect the Arduino to the computer.

---

3. https://www.adafruit.com/products/352
4. https://www.adafruit.com/products/165

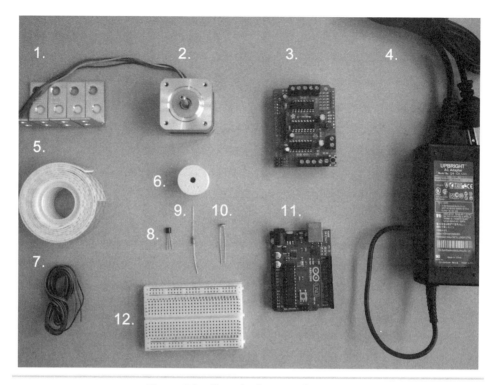

Figure 30—Curtain Automation parts

This project assumes you already have curtains hung on a pulley-based system. If you don't already have curtains in place, there are a number of how-to sites on the Web to assist with hanging curtain rods and setting up the pulley system. The project works best with the continuous drawstring, traverse rod-based hanging curtains. That is, as you pull down on the left side of the string, the right side goes up, and vice versa.

You will also need find the size of pulley wheel that best suits your curtain configuration. For simple curtain rod/drawstring-based systems, a 1-inch-diameter grooved rubber pulley wheel should do just fine. You can obtain a variety of pulley wheel sizes from home hardware, auto parts, and even some craft stores.

Ideally, the center hole of the pulley wheel should snuggly slip onto the stepper motor's drive shaft so that it doesn't fall off or slip when the shaft is rotating. If you have a home improvement store nearby, bring your stepper motor and try the various pulley wheels at the store to save time and hassle. Once you've found the perfect size and fit, you're ready to start assembling the project.

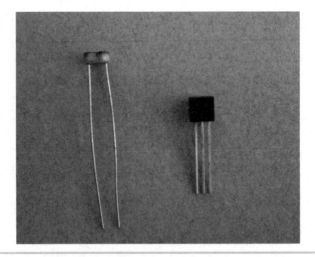

Figure 31—Curtain Automation sensors

8.2 Building the Solution

We have several objectives to complete for our project to work as intended. First, we will test the stepper motor by writing a sketch using Adafruit's AFMotor library. This assumes that you have already constructed the motor shield. Follow the instructions on Adafruit's website for more details on assembling and using the motor shield.[5]

After we have a working stepper motor's drive shaft that rotates back and forth based on the instructions we have the Arduino execute, we will hook up the photo resistor. We will borrow from the same light sensor routine we wrote for the *Tweeting Bird Feeder*. When this photosensor detects light that exceeds the threshold we establish, the stepper motor drive shaft will spin clockwise for a predetermined number of revolutions. When light diminishes below the low threshold value that we set, the shaft will spin the same number of revolutions in the opposite direction. When the shaft spins, the attached pulley will open or close the curtain accordingly.

In addition to light detection, we also need to account for room temperature in case it exceeds a certain value. If the room starts to get too warm, we can spin the drive shaft counterclockwise to close the curtain even when there is daylight. If the temperature cools off to a comfortable level and it's still daylight, the shaft will spin clockwise to reopen the curtains.

---

5. http://www.ladyada.net/make/mshield/make.html

Once our detectors are working and tested, we will attach the pulley wheel to the stepper motor's drive shaft. Then we will wrap the curtain drawstring around it and determine where we can mount the stepper motor/pulley assembly on the wall. The location of the assembly needs to keep the curtain drawstring taut enough so that it will not slip when the pulley is revolving. Next we will calibrate the number of revolutions required by the stepper motor to open and close the curtain. After those settings are determined, we can increase the speed (revolutions per minute) of the stepper motor to establish how quickly or slowly the curtain should open or close.

With these steps in mind, let's first take a look at how we can write a sketch that will control the stepper motor.

8.3 Using the Stepper Motor

Electrical motors work on the principle of electromagnetism to drive the central shafts. As the magnetic field changes around the coils that wrap around the shaft, the change in current propels the shaft forward or backward. Stepper motors refine this principle by allowing granular control of the motor to "step" in well-defined increments. This makes these motors excellent choices for any mechanical task requiring precise control.

Stepper motors are used in ink jet printers, plotters, and disk drives; they are also found in a number of types of industrial manufacturing equipment.

The motor we will use for this project is a popular 12-volt, 350-milliamp, 200-steps per revolution bipolar stepper motor. This motor should provide enough torque to move all but the heaviest of curtains. Because the motor pulls 12 volts of power, it will need to operate from a 12-volt power supply instead of the 5 volts that the standalone Arduino board can deliver. Fortunately, the Arduino board has the electronics necessary to accept a 12-volt power supply that can power the Arduino, the motor shield, and the stepper motor.

Assuming that you have already built a working Adafruit motor shield, here are the steps needed to set up the stepper motor for programming:

1. Connect the four wires from the 12-volt bipolar stepper motor. If your stepper motor is the recommended one from Adafruit, the wiring sequence by color should be red, yellow, green, brown. Refer to the photo in Figure 32, *Bipolar stepper motor wiring*, on page 130.
2. Attach the motor shield to the top of the Arduino Uno.
3. Plug the 12-volt power supply into the Arduino power port.
4. Connect a USB cable from the computer to the Arduino.

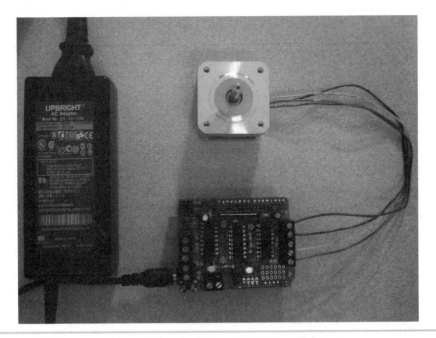

Figure 32—Bipolar stepper motor wiring

Now that the hardware is connected, we can focus on writing an Arduino sketch that will drive the stepper motor.

8.4 Programming the Stepper Motor

In order to get the stepper motor to work the way we want it to, we need to import a library that makes it easy to incrementally rotate the motor's shaft in either direction at the speed we want it to move. Fortunately, controlling a stepper motor is easy thanks to Adafruit's AFMotor motor shield library.[6] As you do with most Arduino libraries, extract the downloaded zip file, rename the extracted folder (AFMotor), and place it in the Arduino libraries folder. For more details, refer to Appendix 1, *Installing Arduino Libraries*, on page 209.

With the AFMotor library installed, launch the Arduino IDE. Let's write a sketch that will test the stepper motor. The code will do the following:

1. Load the AFMotor library.

2. Create an AFMotor stepper motor object and set the stepper's connection and steps per revolution (i.e., how fast the stepper motor's shaft rotates).

---

6. https://github.com/adafruit/Adafruit-Motor-Shield-library

3. Move the shaft clockwise and counterclockwise using the stepper motor's two coils. By the way, this action is known as double-coil activation, and it produces greater torque compared to using just a single coil at a time. We will need that extra torque to move the curtain string.

Here's what the completed sketch should look like:

CurtainAutomation/StepperTest.pde
```
#include <AFMotor.h>

AF_Stepper motor(48, 2);

void setup() {
  Serial.begin(9600);
  Serial.println("Starting stepper motor test...");
  // Use setSpeed to alter speed of rotation
  motor.setSpeed(20);
}

void loop() {
  // step() function
  motor.step(100, FORWARD, DOUBLE);
  motor.step(100, BACKWARD, DOUBLE);

}
```

Note that this test code is essentially a subset of the sample code available from Ladyada's motor shield web page.[7]

Save and upload the sketch to the Arduino. If all goes well, your stepper motor should spin clockwise and counterclockwise until you remove power or upload a new sketch. If the shaft isn't moving, make sure your stepper motor wiring is properly connected. Also make sure that you are using a 12-volt power supply connected to the Arduino, since the motor needs that amount of voltage to move. If you're having a hard time seeing which direction the shaft is rotating, affix a small piece of folded tape on the shaft. It should be easier to see the tape flag move back and forth as the shaft moves.

Now that your hardware is working, it's time to add the temperature and light sensors to give the stepper motor a bit more relevance to its intended motion.

8.5 Adding the Sensors

It's time to combine the working stepper motor with the photosensor we used in the *Tweeting Bird Feeder* project. Photosensor readings will be taken every

---

7. http://www.ladyada.net/make/mshield/use.html

second, and depending on the outdoor light levels, they will trigger the stepper motor to open or close the curtains. We will also add a temperature sensor so that we don't open the curtains if it's already too warm in the room or so that we close the curtains if the room temperature exceeds a predetermined level.

Fortunately for this type of project, which relies on the analog pins to measure light and temperature, the motor shield does not use any of the Arduino's analog pins. Therefore, we will attach one lead of the photocell to the 5V power pin and the other lead to analog pin 0. And just like the Tweeting Bird Feeder project, we need to bridge the 10k ohm resistor from the analog pin 0 to ground. Using a breadboard for this is much easier than wrapping the leads in series. Plus, the breadboard will make a good stand to keep the photocell propped up and angled toward the outdoor light.

The temperature sensor has three leads: the first will connect to the 5V power pin, the temperature sensor's middle lead will connect to analog pin 5, and the third (far right) lead will connect to the ground pin. The motor shield makes this easier. Refer to Figure 33, *Curtain Automation stepper motor and sensor wiring diagram*, on page 133, for setting this up. Note that although the diagram shows an Arduino, the wiring will actually be connecting to the motor shield mounted on top of the Arduino (hence the wiring on the side for the stepper motor connection as well as the wiring to analog pin 5 and the ground and 5V pins on the right side of the shield).

We will poll the variable values of these two sensors every second and react accordingly should the threshold values we established for these measurements be exceeded. Let's write the sketch that will do just that.

8.6 Writing the Sketch

The sketch for this project borrows from ideas we encoded in two other projects. The sensor readings come from the Tweeting Bird Feeder, and the state machine for the open or closed status was copied from the Water Level Notifier. As such, here is the complete sketch.

CurtainAutomation/CurtainAutomation.pde
```
① #include <AFMotor.h>
②  #define LIGHT_PIN          0
   #define LIGHT_THRESHOLD  800
   #define TEMP_PIN           5
   #define TEMP_THRESHOLD    72
   #define TEMP_VOLTAGE     5.0
   #define ONBOARD_LED       13
```

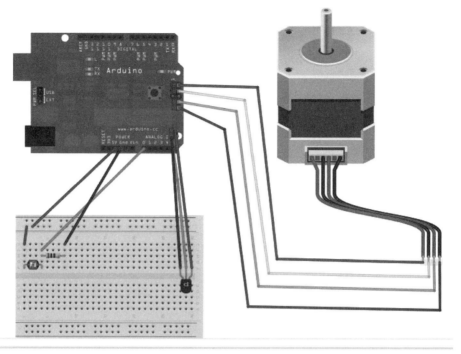

Figure 33—Curtain Automation stepper motor and sensor wiring diagram

```
③ int curtain_state = 1;
   int light_status  = 0;
   double temp_status = 0;

   boolean daylight = true;
   boolean warm     = false;

   AF_Stepper motor(100, 2);

④ void setup() {
     Serial.begin(9600);
     Serial.println("Setting up Curtain Automation...");
     // Set stepper motor rotation speed to 100 RPMs
     motor.setSpeed(100);
     // Initialize motor
     // motor.step(100, FORWARD, SINGLE);
     // motor.release();
     delay(1000);
   }

⑤ void Curtain(boolean curtain_state) {
     digitalWrite(ONBOARD_LED, curtain_state ? HIGH : LOW);
```

```
  if (curtain_state) {
    Serial.println("Opening curtain...");
    // Try SINGLE, DOUBLE, INTERLEAVE or MICROSTOP
    motor.step(800, FORWARD, SINGLE);
  } else {
    Serial.println("Closing curtain...");
    motor.step(800, BACKWARD, SINGLE);
  }
}

❻ void loop() {

  // poll photocell value
  light_status = analogRead(LIGHT_PIN);
  delay(500);

  // print light_status value to the serial port
  Serial.print("Photocell value = ");
  Serial.println(light_status);
  Serial.println("");

  // poll temperature
  int temp_reading = analogRead(TEMP_PIN);
  delay(500);

  // convert voltage to temp in Celsius and Fahrenheit
  float voltage = temp_reading * TEMP_VOLTAGE / 1024.0;
  float temp_Celsius = (voltage - 0.5) * 100 ;
  float temp_Fahrenheit = (temp_Celsius * 9 / 5) + 32;
  // print temp_status value to the serial port
  Serial.print("Temperature value (Celsius) = ");
  Serial.println(temp_Celsius);
  Serial.print("Temperature value (Fahrenheit) = ");
  Serial.println(temp_Fahrenheit);
  Serial.println("");

  if (light_status > LIGHT_THRESHOLD)
      daylight = true;
  else
      daylight = false;

  if (temp_Fahrenheit > TEMP_THRESHOLD)
      warm = true;
  else
      warm = false;

  switch (curtain_state)
  {
  case 0:
      if (daylight && !warm)
```

```
      // open curtain
      {
        curtain_state = 1;
        Curtain(curtain_state);
      }
      break;

  case 1:
      if (!daylight || warm)
      // close curtain
      {
        curtain_state = 0;
        Curtain(curtain_state);
      }
      break;
  }
}
```

① Reference the AFMotor library that will be used to drive the stepper motor attached to the Adafruit motor shield.

② We will define several values up front. This will make it easier to change the LIGHT_THRESHOLD and TEMP_THRESHOLD values as we refine the stepper motor trigger points.

③ Variables for storing curtain state—as well as the analog values of the photocell and temperature sensors two boolean variables, daylight and warm —are used in the main loop's conditional statements to identify the status of daylight and the indoor room temperature. We also assign the number of steps per revolution (in this case, 100) and the motor shield port that the stepper motor is attached to (in this case, the second port per the wiring diagram) by creating an AF_Stepper object called motor.

④ Here's where we initialize the serial port to output the light and temperature readings to the Arduino IDE serial window, as well as initialize the speed of the motor (in this case, 100 revolutions per minute).

⑤ The Curtain function will be called when the light or temperature thresholds are exceeded. The state of the curtains (open or closed) is maintained so that the motor doesn't keep running every second the threshold is exceeded. After all, once the curtains are opened, there's no need to open them again. In fact, doing so might even damage the stepper motor, grooved pulley, or curtain drawstring.

If the Curtain function receives a curtain_state of true, the stepper motor will spin counterclockwise to open the curtains. A curtain_state value of false will spin clockwise to close the curtains.

We will also use the Arduino's onboard LED to indicate the status of the curtains. If the curtains are open, the LED will remain lit. Otherwise, the LED will be off. Since the motor shield will be covering the top of the Arduino, the onboard LED won't be easily visible, but it will still serve as a good visual aid for debugging purposes.

⑥ The main loop of the sketch is where all the action happens. We poll the analog values of the photocell and temperature every second, convert the electrical value of the temperature sensor both to Celsius and—for those who have yet to convert to the metric system—Fahrenheit. If the light sensor exceeds the LIGHT_THRESHOLD value we assigned in the #define section of the sketch, then it must be daytime (i.e., daytime = true). However, we don't want to open the curtains if it's already warm in the room, since the incoming sunlight would make the room even warmer. Thus, if the temperature status exceeds the TEMP_THRESHOLD, we will keep the curtains closed until the room cools down. After checking the status of the curtain_state, we will pass a new state to the Curtain routine and open or close the curtains accordingly.

Verify, download, and execute the sketch on the Arduino. Leave the Arduino tethered to your computer and open the Arduino IDE's serial window to see the light and temperature values being captured by the sensors. Now we can verify whether exceeding the threshold values produces the desired effect of activating the stepper motor (see Figure 34, *Test the Curtain Automation sketch*, on page 137).

Test the Curtain Automation sketch first by covering the photocell with your finger to verify that the stepper motor rotates the shaft in the counterclockwise direction. Remove your finger, and the shaft should rotate clockwise the same number of times. Blow warm air or use a blow dryer to warm up the air around the temperature sensor. When the threshold is exceeded, the stepper motor shaft should spin clockwise. This translates to open curtains being closed. Before removing the heat source, cover the photocell with your finger again. Then remove the heat source. The stepper motor shaft should remain motionless.

Remove your finger from the photocell. If the air surrounding the temperature sensor has cooled, the shaft should rotate counterclockwise. If it doesn't rotate, blow air on the temperature sensor to cool it down; it should react once the temperature drops below the target threshold. Verify that your sketch reacted when the designated threshold values for light and temperature were exceeded. You may need to tweak these threshold values to ensure that the stepper motor reacts when the desired light densities and room temperature are

Figure 34—Test the Curtain Automation sketch

attained. You may also need to consider omitting a light or temperature sensor range for the assigned threshold values. Otherwise, the stepper motor may act a bit jittery as the light or temperature wavers back and forth between triggered threshold values.

After you have confirmed that the sensors are properly reporting their values and the stepper motor shaft moves when the threshold values are exceeded, we can seat the sensors on a windowsill and mount the stepper motor on the wall next to the curtains. Once the system is working, you can also choose to reposition the sensors anywhere inside the room as long as the wire attaching the sensors to the Arduino board is long enough. If you do so, make sure that the wire and sensors are not in an area where they might accidentally be stepped on or where the connecting wire could be tripped over.

8.7 Installing the Hardware

When setting up the sensors, you can leave them seated in the small breadboard we used during the testing of the Curtain Automation sketch. I used a piece of foam double-sided tape to keep the breadboard seated in place, with the sensors pointing toward the window like a high-tech flower box. Also, the stepper motor tends to get very warm when in use, so as an added safety precaution, be sure to mount the motor away from anything flammable. For example, make sure the stepper motor is mounted away from the curtains or shade that you're opening and closing!

Measure the distance from the breadboard sitting on the windowsill to where you want to place the Arduino+motor shield. The Arduino can be mounted on a table, in an enclosure, or even on the wall if you prefer. I recommend using an extra foot or two of wire wrapped in a loop just in case you need to relocate the Arduino later. Also take into account the placement of the 12V power supply brick and the electric cord that has to plug into the Arduino to power the Arduino, the motor shield, and the stepper motor.

Slip the rubber-grooved pulley wheel onto the stepper motor shaft. Loop the curtain drawstring around the pulley wheel. Pull the stepper motor down until the drawstring is taut around the pulley wheel. Before permanently mounting the stepper motor, attach the four angle brackets to it using double-sided foam tape. The tape will keep the motor in place while you screw in the mounting brackets. The foam tape will also help dampen vibrations against the wall and keep the operation quiet when the stepper motor shaft is operating. You may also want to affix the stepper motor first to the wall, using tape to hold it in place, just to be sure that the curtain drawstring isn't looped too tightly or too loosely around the pulley attached to the stepper motor shaft. Don't make the string too taut in case you need just enough slack to allow for recalibration should the string happen to slip when the pulley spins.

Perform a few tests before screwing the angle brackets to the wall. This will verify that the drawstring around the pulley has just the right amount of tension and friction to be pulled by the rotating pulley when the stepper motor runs. When you're satisfied with the placement of the stepper motor, screw the four angle brackets into the wall.

Calibrate the speed and number of revolutions that the stepper motor needs to make to fully open and close the curtains. Start in small increments at first, remembering to apply the same number of revolutions in both clockwise and counterclockwise directions. You can estimate the number of revolutions needed to draw the curtains open and closed by measuring the distance that the drawstring moves with each revolution of the pulley. Divide this by the total distance that the curtains need to move to completely open and close. This will give you the total number of stepper motor shaft rotations you need to program to open and close the curtains.

```
Distance curtain moves with one stepper motor shaft rotation = 5 centimeters
Total distance curtain needs to move from start to finish = 90 centimeters
90 cm / 5 cm = 18 rotations
```

Figure 35—Curtain Automation, installed and calibrated

When the system is perfectly calibrated, mark the drawstring with a felt marker at the points where the string meets the pulley when the curtains are opened and closed. This will help should you need to recalibrate the drawstring if it falls out of sync over time. Once configured, your setup may look like Figure 35, *Curtain Automation, installed and calibrated*, on page 139.

Test the system a few times by covering the photocell and artificially heating the temperature sensor with your breath or a blow dryer. Observe when the photocell triggers the curtain opening and closing events. If it's too sensitive or opens the curtains as a result of indoor light reflecting off the glass, you may need to reposition the photocell in a different location. I taped the sensor to a corner of the window using black electrical tape. This helped minimize the sensor from being exposed and reacting to indoor room lighting.

Allow the assembly to run a few days, noting when the curtains should react to light or temperature triggering the sensors. Alter the temperature and light sensitivity values accordingly. Once everything is set up correctly, you should only need to check on the curtain string's position once every couple of weeks for any recalibration adjustments. After a while, you will simply take the autonomous curtains for granted. Visiting guests seeing the curtains in operation for the first time will be amazed by your high-tech handiwork.

8.8 Next Steps

The motor shield can handle up to two stepper motors at a time. This might be useful in a large room with more than one window. The curtains' pulley systems can also be linked together so that a single stepper motor could open and close multiple curtains/shades. Going even further, the stepper motor/motor shield combination can be employed in a number of other home automation scenarios.

- Elaborate on the curtain pulley with more sophisticated curtain rod systems that have an interior up/down window blind and exterior left/right curtain draw. Time one motor to raise/lower the blind followed by opening/closing the decorative exterior curtain. Modify these two configurations based on heat and light (i.e., if hot daylight, open curtain but close blind).

- Add a PIR sensor to the Arduino+motor shield assembly so you can open and close the curtains when motion is detected in the room.

- Network-enable the curtain automation assembly with an onboard LED or an Arduino Ethernet so you can open and close your curtains from your smartphone or write a script that will operate the curtains during certain times of the day.

- Keep the Arduino with motor shield assembly connected to the USB port of your computer and drive the curtains remotely via USB-to-serial communication. Set up a web application server to expose the open/close methods as web services to be called from a native smartphone application. Write a script that runs on the host computer to open and close the curtains at a specified date and time.

- Repurpose the pulley system for Halloween fun by swapping out the light and temperature sensors with a motion sensor. Attach the pulley string to big paper spiders that go up and down when motion is detected.

- Create a stepper motor-driven carousel for clothing with unique RFID tags affixed to each hanger. Queue up a random outfit or base the selection on criteria such as day, month/season, and current outdoor temperature. Make a phone or tablet app that allows you to gesture to a photo of your desired outfit onscreen and then have that clothing selection waiting for you front and center in your closet.

CHAPTER 9

Android Door Lock

Tired of carrying around old-fashioned metal keys to your home? You're probably already carrying a smartphone. Wouldn't it be much more convenient to open your front door via an app that you built for your smartphone instead? Wouldn't it also be a nice security feature to take a photo of the person(s) unlocking your door with this app and email that captured photo as an attachment to yourself? (See Figure 36, *Open doors wirelessly using a smartphone*, on page 142.)

In this project, we are going to use an inexpensive, first generation Android phone. We will connect it to a Sparkfun IOIO ("yo-yo") board and a relay switch to operate an electrified door latch. The first-gen Android phone will run a server that will respond to your unlock requests sent from a second Android phone running the door unlock client. When the unlock request is triggered, the server phone will snap a photo using the phone's built-in camera and silently email the captured image to you. Let's go and make it!

9.1 What You Need

I originally designed this project using a relay switch that we would have constructed part by part. This relay would have been used to turn on and off the power to the electric door latch. But after reviewing the potential safety hazards associated with improper wiring and handling of the circuit, I decided to take a safer, more conservative approach.

Instead of wrestling with the potentially jolting perils of accidental relay shocks, we will use a product specifically designed to address these concerns. Called the PowerSwitch Tail II, this simple switch houses a relay that can control standard 120V electrical devices. The relay can be energized via a 5V signal from a digital pin of a microcontroller board such as an Arduino or, in

Figure 36—Open doors wirelessly using a smartphone

the case of this project, the PIC-based IOIO board. The PowerSwitch Tail's prebuilt relay circuit is far easier and safer than building your own, and the cost is quite reasonable compared to the expense of procuring and assembling these parts on your own.

Rather than using an Arduino connected to a computer for data processing and control, we are going to use an Android phone connected to Sparkfun's IOIO board. This hardware combination will serve the same function as an Arduino/PC coupling but without the size, bulk, and energy requirements that an always-on Arduino/PC combination would entail.

So what exactly is an IOIO board? It is a hardware bridge that allows Android phones to communicate with whatever sensors and motors are connected to the board. The IOIO board connects to the phone via Android's USB debugging pathway. This pathway can be used to send and receive signals to and from the IOIO's onboard PIC processor.

IOIO's designer, Google software engineer Ytai Ben-Tsvi, designed the IOIO prior to Google's official Open Accessory Protocol (ADK) initiative,[1] but he is

---

1. http://accessories.android.com

working to make the board fully compatible with the ADK specification. The ADK is part of Google's Android@Home home automation initiative. Investing in the board not only gives you the tools you need to make it work today, but it will also play nice with the future Android@Home APIs. And even more importantly, the IOIO works especially well today for custom projects like the one we will build.

Here is a list of all the parts we will need to construct the Android Door Lock (refer to the photo in Figure 37, *Android Door Lock parts (some preassembled)*, on page 144):

1. A PowerSwitch Tail II (PN 80135)[2]
2. A 2.1mm female barrel jack cable to safely connect the 12V power supply to the electric door strike[3]
3. A 5VDC 1A power supply[4]
4. A 12V 5A switching power supply to electrify the electric door strike[5]
5. Three pieces of wire
6. An Android OS smartphone with a built-in camera, preferably the original Android G1 phone. This phone can be purchased from sites like craigslist.org or ebay.com for under $100 US. Note that not all Android phones are compatible with the IOIO Board. Check the IOIO Board discussion group for more details.[6]
7. A barrel jack to 2-pin JST cable that will connect to the IOIO board's JST right angle connector[7]
8. A Sparkfun IOIO Board with JST right angle connector[8]
9. A Smarthome Electric 12VDC Door Strike[9]
10. A standard A to Mini-B USB cable to connect the G1 Android phone to the USB port on the IOIO board

You will also need a second Android device (phone, tablet, etc.) that can run the Door Lock client application along with the Eclipse IDE, Android SDK 1.5 or higher, and the Android Development Tools (ADK) plugin for Eclipse. Refer to the *Web-Enabled Light Switch* project for more details about the Android development requirements.

---

2. http://www.sparkfun.com/products/10747
3. http://www.adafruit.com/products/327
4. http://www.sparkfun.com/products/8269
5. http://www.adafruit.com/products/352
6. https://groups.google.com/group/ioio-users?pli=1
7. http://www.sparkfun.com/products/8734
8. http://www.sparkfun.com/products/10748 and http://www.sparkfun.com/products/8612, respectively.
9. http://www.smarthome.com/5192/Electric-Door-Strike-Mortise-Type/p.aspx

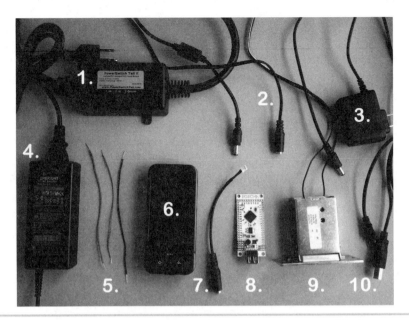

Figure 37—Android Door Lock parts (some preassembled)

While this project is one of the most expensive to build in this book, it is also one of the most flexible in terms of reusing and extending the hardware investment. Once you become familiar with how to leverage the IOIO board with an Android phone, you will understand why Google is so enthusiastic about their Android@Home effort. You will also learn how you can easily create a new category of home automation applications implementing your own ideas. But before you can attain those greater heights of actualization, you have to understand the basics. That's what we're going to do in the next section.

9.2 Building the Solution

This is a fairly complex project and certainly the most challenging one in the book. We will spend a majority of the time assembling and testing this project's hardware. After the hardware has been tested, we'll program the Android phone to talk to the IOIO board, the onboard phone's camera, and the wireless network. Then we'll write a simple Android client application we can execute from another Android device that will trigger the IOIO board to turn on the PowerSwitch Tail, which will in turn power the electric door strike that unlocks the door. Here are the steps we'll take to assemble, program, and deploy the Android Door Lock:

> **IOIO Successor?**
>
> Sparkfun announced a new, ADK-compliant development board based on the Arduino Mega ADK that uses Google's Open Accessory protocol. Called the Electric Sheep,[a] this new board also features an onboard FTDI header and DC power connector, obviating the need for the separate components required by the IOIO. However, the board is also twice as expensive as the IOIO and requires more power. Fortunately, the Electric Sheep has plenty of analog and digital pins and can be programmed using Google's ADK and the Arduino Android Handbag.[b] As such, this more powerful Android@Home-friendly board will likely replace the IOIO board in the future.
>
> ---
>
> a. http://www.sparkfun.com/products/10745
> b. http://handbagdevices.com/

1. Attach the JST connector to the Sparkfun IOIO board so that the IOIO board can be powered by the 5V power supply.

2. Tether the Android G1 phone to the IOIO board via the USB cable.

3. Plug the Smarthome Electric 12VDC Door Strike into the 12V power supply via the 2.1mm female barrel jack cable.

4. Connect the PowerSwitch Tail to the IOIO board via three wires for the PowerSwitch Tail's power, control, and ground connectors.

5. Program the Android phone to trigger the PowerSwitch Tail via the IOIO board.

6. Snap a photo using the Android phone's built-in camera when the PowerSwitch Tail is triggered.

7. Send the resulting image as a message attachment to a designated email recipient.

8. Write a native client application for a second Android device that will be used to unlock the door strike.

9. Install the Electric Door Strike in the desired doorframe, routing the electrical wiring to a nearby outlet.

10. Bundle the controller components (the PowerSwitch Tail circuit and the IOIO board) into an easily accessible wall-mounted lock box that can be serviced in case parts need to be replaced.

11. Mount the Android phone running the door lock server application near the entry with the camera lens facing the door so images of those entering can be easily captured.

The most time-consuming aspects are the soldering of the JST right angle connector to the IOIO board and the wiring of the circuit between the IOIO board and the PowerSwitch Tail. Everything else is simply a matter of plugging into the right segment in the series. Essentially, the 5V power supply plugs into the IOIO board, the Android phone plugs into the IOIO via the USB cable, the PowerSwitch Tail, controlled by the IOIO board, plugs into the wall on one side and the 12V power supply on the other, and the electric door strike plugs into the 12V power supply.

The first thing you should do is solder the JST right angle connector to the IOIO board. It would have been much easier for Sparkfun customers had the connector been preinstalled on the board. But hey, that's part of the fun (along with the terror should something go horribly wrong when an expensive component accidentally gets fried) that these projects have to offer. Fortunately, soldering the connector isn't too difficult and will make powering the board and the Android phone vastly easier.

After the JST connector is attached to the IOIO board, plug the barrel jack to 2-pin JST cable into the JST connector on the board and the 5V power supply. Then attach the Android phone via a USB cable to the IOIO board. You're halfway there!

Attach the positive and negative wires of the electric door strike to the positive and negative leads of the 2.1mm female barrel jack cable. The positive wires are those with a white strip along the side of the wire casing. Use electrical tape or, better still, heat shrink tubing to safely cover any exposed wire. Connect the barrel jack to the 12V power supply.

Test the strike by plugging the 12V power supply into your standard 120V wall socket. You should see the LED on the power supply light up, followed by an audible click from the strike. Note that while electrified, you will be able to move the spring latch on the strike back and forth without much resistance. Unplug the 12V power supply from the wall and the strike will return to its fixed, nonelectrified state. Consequently, you should be unable to move the latch.

Next, let's connect the PowerSwitch Tail to the IOIO board. Follow along with Figure 38, *Android Door Lock wiring diagram*, on page 147. Using three wires, connect one from the ground pin on the IOIO board to the the negative lead on the PowerSwitch Tail.

Connect the middle (control) lead on the PowerSwitch Tail to the IOIO board's digital pin 3. Why not pin 0, 1, or 2? That's because not all IOIO boards can handle the 5V signal required to electrify the relay in the PowerSwitch Tail.

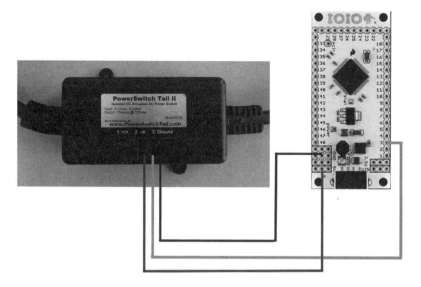

Figure 38—Android Door Lock wiring diagram

Pulling 5V from a pin not capable of this voltage could damage the IOIO board. Refer to the IOIO board wiki or simply flip the IOIO board over and look for the pins that are enclosed by a white circle.[12] The circle indicates that the enclosed pin is 5V pullup capable. When wired, your IOIO board should look like the one shown in Figure 39, *Wiring the IOIO board*, on page 148.

Finally, attach a wire from the positive lead from the PowerSwitch Tail to any of the three 5V pins on the left lower corner of the IOIO board. The circuit is complete.

At this point, nothing will happen until we program the necessary instructions to turn pin 3 on and off, thereby signaling the PowerSwitch Tail to do the same. As such, we are going to write a simple Android program with an onscreen toggle switch that will instruct pin 3 to do just that.

9.3 Controlling the Android Door Lock

Before we can write an elaborate server for the Android phone, we first need to write a test program that will validate the circuit we constructed in the last section.

---

12. https://github.com/ytai/ioio/wiki

Figure 39—Wiring the IOIO board

Be sure to have installed the plug-ins for the Android SDK and Eclipse IDE with the Android Development Tools on your computer. Refer to Chapter 7, *Web-Enabled Light Switch*, on page 105, for the steps on how to configure the SDK and IDE if you haven't already done so. Then, download the HelloIOIO demo project from the Sparkfun IOIO tutorial web page.[13] The HelloIOIO project is a simple application that turns the IOIO's onboard LED on and off. We are going modify this simple application by declaring another ToggleButton object in its main.xml layout file. Then we'll add four lines of code to the MainActivity.java file that describe the added ToggleButton action for digital pin 3 on the IOIO.

Import Sparkfun's HelloIOIO project into the Eclipse environment via the File →Import→Existing Projects option. If you prefer, you can also load the modified HelloIOIO-PTS project available from this book's code download page that has all the necessary code additions mentioned in this section.

All IOIO board projects rely on the custom IOIOLib library that must be added to each IOIO project. Use the following steps to do so:

1. Import the IOIOLib bundle into the Eclipse environment via the same File →Import→Existing Projects into Workspace menu option.

2. Highlight the HelloIOIO project in the Eclipse Package Explorer pane.

3. Access the Properties option from the Eclipse Project menu.

---

13. http://www.sparkfun.com/tutorials/280

4. Select Android from the selections on the left column of the Properties dialog box.

5. Click the Add... button. A Project Selection dialog box should pop up listing the IOIOLib project. Highlight the IOIOLib item and click the OK button.

If IOIOLib was successfully imported, it should be listed with a green checkmark in the Library portion of the Properties dialog box, as shown in Figure 40, *IOIOLib successfully imported and referenced*, on page 150.

With the IOIOLib properly referenced, edit the /res/layout/main.xml file from the HelloIOIO project. Add another `ToggleButton` object to the existing layout by copying the existing `TextView` description containing the `ToggleButton` description of the toggle button used to turn on and off the IOIO's onboard LED. Paste it in right after the original `TextView` section. Then, rename the android/id value of the copied toggle button to `android:id="@+id/powertailbutton"`. This will be the reference accessed in the modified `MainActivity` class. The modified main.xml file should look like this:

AndroidDoorLock/HelloIOIO-PTS/res/layout/main.xml
```xml
<?xml version="1.0" encoding="utf-8"?>
<LinearLayout xmlns:android="http://schemas.android.com/apk/res/android"
    android:orientation="vertical"
    android:layout_width="fill_parent"
    android:layout_height="fill_parent">
<TextView
    android:layout_width="fill_parent"
    android:layout_height="wrap_content"
    android:text="@string/txtLED"
    android:id="@+id/title"/>
<ToggleButton android:text="ToggleButton"
    android:layout_width="wrap_content"
    android:layout_height="wrap_content"
    android:id="@+id/button">
</ToggleButton>
<TextView
    android:layout_width="fill_parent"
    android:layout_height="wrap_content"
    android:text="@string/txtPowerTail"
    android:id="@+id/title"/>
<ToggleButton android:text="ToggleButton"
    android:layout_width="wrap_content"
    android:layout_height="wrap_content"
    android:id="@+id/powertailbutton">
</ToggleButton>
</LinearLayout>
```

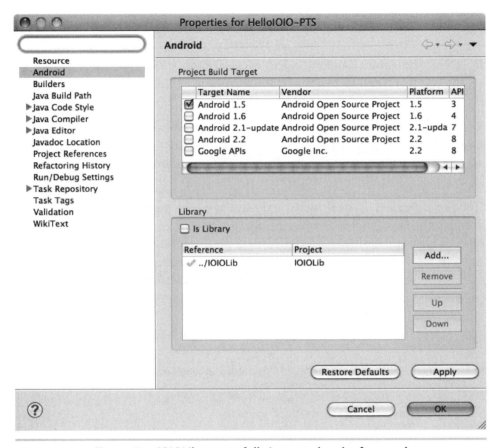

Figure 40—IOIOLib successfully imported and referenced

Next, add the code for the second toggle button to the MainActivity class that will turn on and off the signal going to the PowerSwitch Tail. The first addition is the powertailbutton_ = (ToggleButton) findViewById(R.id.powertailbutton); line, which associates the powertailbutton_ object with the powertailbutton toggle button defined in the main.xml file.

With the user interface addition of the toggle button for the PowerSwitch Tail, we can add the object reference to the MainActivity class located in the /src/ioio/examples/hello/pts/MainActivity.java file:

AndroidDoorLock/HelloIOIO-PTS/src/ioio/examples/hello/pts/MainActivity.java
```
private ToggleButton button_;
private ToggleButton powertailbutton_;
```

Instantiate the powertailbutton_ object in the MainActivity OnCreate method, like this:

AndroidDoorLock/HelloIOIO-PTS/src/ioio/examples/hello/pts/MainActivity.java
```
@Override
public void onCreate(Bundle savedInstanceState) {
    super.onCreate(savedInstanceState);
    setContentView(R.layout.main);
    button_ = (ToggleButton) findViewById(R.id.button);
    powertailbutton_ = (ToggleButton) findViewById(R.id.powertailbutton);
}
```

So when the main application window is created, the toggle button for the PowerSwitch Tail will now be accessible via the MainActivity class. All that remains is the code needed to listen for the powertailbutton_ toggle action being turned on and off:

AndroidDoorLock/HelloIOIO-PTS/src/ioio/examples/hello/pts/MainActivity.java
```
class IOIOThread extends AbstractIOIOActivity.IOIOThread {
    /** The on-board LED. */
    private DigitalOutput led_;
①   private DigitalOutput powertail_;
    /**
     * Called every time a connection with IOIO has been established.
     * Typically used to open pins.
     *
     * @throws ConnectionLostException
     *             When IOIO connection is lost.
     * @see ioio.lib.util.AbstractIOIOActivity.IOIOThread#setup()
     */
    @Override
    protected void setup() throws ConnectionLostException {
        led_ = ioio_.openDigitalOutput(0, true);
②       powertail_ = ioio_.openDigitalOutput(3,true);
    }
    /**
     * Called repetitively while the IOIO is connected.
     *
     * @throws ConnectionLostException
     *             When IOIO connection is lost.
     *
     * @see ioio.lib.util.AbstractIOIOActivity.IOIOThread#loop()
     */
    @Override
    protected void loop() throws ConnectionLostException {
        led_.write(!button_.isChecked());
③       powertail_.write(!powertailbutton_.isChecked());
        try {
            sleep(10);
        } catch (InterruptedException e) {
        }
    }
}
```

① Initialize the DigitalOutput powertail_ object.

② Assign the powertail_ object to the IOIO's digital pin out 3.

③ Turn on or off the digital signal (i.e., make it High or Low) to IOIO's digital pin out 3 when the onscreen toggle button for the PowerSwitch Tail is toggled on or off.

When the onscreen PowerSwitch Tail toggle button is switched on, it will instruct the powertailbutton_ instance to send a 5V signal from digital pin 3. This in turn will electrify the PowerSwitch Tail relay to power on, which will then electrify the 12V power adapter that will ultimately electrify and release the lock.

Save your changes, compile the Android application, and install the modified HelloIOIO program on the phone. Check to ensure that your door hardware circuit is properly wired and powered. Then plug in the USB cable between the phone and the IOIO board and execute the modified HelloIOIO program on the phone.

If nothing happens, verify that the USB Debugging option is checked on the phone. Also, make sure your wiring is connected correctly. If you have access to a multimeter or an oscilloscope, check to see that 5 volts are flowing from the digital pin 3 when the onscreen PowerSwitch Tail toggle switch is set to the on position. If the output is less than 5 volts, there will not be enough of a signal from the IOIO board to electrify the PowerSwitch Tail and thus power the electric door latch.

Now that the hardware is working properly, we will network-enable the lock so we can open it by requesting a URL from a web server that we will add to this modified HelloIOIO program.

9.4 Writing the Android Server

Time to write the door lock server. Instead of relying on a personal computer to perform the heavy lifting of running Python scripts to respond to incoming requests, we are going to use the computing power embedded in the Android smartphone itself. Even older Android phones are computationally more powerful than desktop computers were only a few years before the Android OS was introduced.

Besides, an Android phone acting as this project's server offers a number of advantages:

- Power requirements are far lower than a desktop computer, making for much greener energy consumption.

- The phone has onboard Wi-Fi, allowing it to be placed anywhere within range of the home's wireless access point.

- The phone has an onboard camera that can be programmed using standard SDK calls to capture images.

- The phone has other features like Bluetooth and speech synthesis that we will be using in the *Giving Your Home a Voice* project.

The phone-based web server application will need to perform the following functions:

1. Establish a standard web server instance and listen for inbound requests for a specific URL.

2. When the URL is requested, send a signal for five seconds to power IOIO board pin 3. This will release the electric lock long enough to allow entry.

3. After the five seconds, use the built-in camera on the web server host device to take a photo of the person entering the door.

4. Send the captured image as an email attachment to a designated recipient.

5. Return to an idle state and await another properly formatted inbound request to begin the cycle anew.

In order to construct the web server, we will borrow code snippets from the open source GNU GPLv3 Android Web Server project available on Google Code.[14] We will also incorporate code (generously posted by Jon Simon) for sending email messages with attachments from an Android application without having to rely on intents to do so.[15]

Since most intents typically rely on user interaction, it won't work for our standalone web server scenario. Combining these two projects with the IOIO code will allow our program to autonomously listen for and react to door unlock requests. Lastly, we will rely on bits of Camera Sample code written by Krishnaraj Varma to capture an image and save it on the Android's SD card.[17] It will be this image that we will send as an email attachment. However, before we can start working on this Android program mashup, we need a more definitive way to access the IP address of the Android phone.

---

14. http://code.google.com/p/android-webserver/
15. http://www.jondev.net/articles/Sending_Emails_without_User_Intervention_%28no_Intents%29_in_Android
17. http://code.google.com/p/krvarma-android-samples/

> **Joe asks:**
> ### What Is an Android Intent?
>
> According to the Android developer documentation, an intent is "an abstract description of an operation to be performed."[a] In layman's terms, intents are used to transmit and receive messages between Android activities and services. Intents can also send messages to the same application that generates them, though intents are more frequently used to send a message from one application, say a web browser that has downloaded an audio file, to another application, such as a music program.
>
> When multiple applications have been registered to receive certain Intent messages, a pop-up dialog box might appear, asking the user to select which application to send the message to. If you're an experienced Android device user, you have no doubt seen this pop-up appear at one time or another. Android allows users to select via a checkbox in the pop-up dialog the default application to send such messages to so as not to annoy you with chronic pop-ups all the time.
>
> As a result of such user interaction requirements, intents are rarely optimal for entirely autonomous operations, such as sending email, since the message receiving the Intent message (in this case, an email application) might still require user interaction to complete the intended action (i.e., the user would need to click the Send button in the email program to actually send the email message initiated by the original Intent-transmitting program).
>
> ---
> a. http://developer.android.com/reference/android/content/Intent.html

Setting a Static IP Address

By converting the phone's Wi-Fi IP address from a dynamic to a static address, it will be much easier to repeatedly locate the phone on a home wireless local area network. If you haven't already created a static IP range on your wireless router, either do so or set the IP address to something higher than 200, since it's unlikely you will have that many devices requesting an IP address from the DHCP server in your wireless router anytime soon.

You can access the configuration setting on most Android phones by selecting the Settings icon, followed by the Wireless and Network menu selection. Then select Wi-Fi settings and press the menu button on the Android phone itself to bring up a pop-up menu with Scan and Advanced selections. Click the Advanced menu option. You will then see a screen of menu choices allowing you to modify a number of network settings, one of which is a Use static IP checkbox. Click that checkbox to enable the ability to set the Wi-Fi radio's IP, subnet, and gateway address as shown in Figure 41, *Configuring an Android device to use a static IP*, on page 156. Set these according to your wireless router network configuration. For example, if your wireless network router is leasing an address range beginning at 192.168.1.2, your settings can most likely be configured to the following:

- IP Address: 192.168.1.230
- Gateway: 192.168.1.1
- Netmask: 255.255.255.0

Set the DNS1 and DNS2 values to the DNS address of your choice (I used Google's Public DNS in my configuration), though it's best to set these addresses to the same domain name servers that your other network clients are using to maintain consistency on your local area network. When you have entered the static values, click the Menu button on your Android device and select the Save option.

Test access to the static IP on the phone by pinging it from another computer on your network. If you set up the static IP address information successfully, you should see positive ping results. If not, check your settings and be sure to save your changes. With the static IP address confirmed, we're ready to proceed with writing and testing some Android web server code.

Creating an Android Web Server

Android runs a modified version of the Java Virtual Machine and as such, brings to it a number of standard Java libraries. That's a good thing, since one of the libraries helps to make creating and running a web server trivial by using just a few lines of code.

Rather than taking up book space showing the contents of the dozens of files that comprise the full program listing, visit the book's website and download the DoorLockServer.zip file. Once downloaded and uncompressed, import the project into your Android SDK-configured Eclipse environment via the File->Import... menu option. If you examine the file's contents, you will notice a file named AndroidDoorLockServerActivity.java. Look for the two lines of code in the private void startServer(int port) method that uses the Android phone's Wi-Fi IP address, port number, and default message handler to start the web server on the phone.

```
server = new Server(ipAddress,port,mHandler);
server.start();
```

This instruction imports the ServerSocket reference and tells Android to listen for requests on port 80 on our assigned static IP address. Naturally, there is much more to manage, such as starting and stopping the server from the UI, making the server a service so Android keeps it running in the background,

Figure 41—Configuring an Android device to use a static IP

keeping the phone from entering sleep mode, acting on inbound requests, and handling errors.

Now that we have the basic requirements for running a web server from an Android device, the next task we need to tackle is to combine it with the IOIO board functionality we enabled in Section 9.3, *Controlling the Android Door Lock*, on page 147.

Web Server + IOIO Board

This is where things get interesting. By combining the IOIO test application we wrote in Section 9.3, *Controlling the Android Door Lock*, on page 147, with the web server in the last section, an inbound HTTP request will trigger digital pin 3 on the IOIO board. This will signal the PowerSwitch Tail to allow power to go to the electric door strike. Essentially, we will transplant the IOIO trigger routine into the web server's response to an HTTP request. For example, calling a URL like *http://192.168.1.230* will ultimately energize the door lock and allow entry.

We don't want to leave the door permanently unlocked by keeping digital pin 3 on, so we will have to turn off power after a set amount of time. Five seconds should be adequate for our testing purposes. To do so, we will call upon Android's Thread.sleep() function to pause program execution for a set duration.

Experienced Android application developers know that this isn't the most elegant way to handle pausing program execution because it can make user interface elements appear unresponsive. However, since the Android device will be used as a server rather than a client, we won't have to worry too much about optimizing the interactive user experience for this program. I set the delay to five seconds (Thread.sleep(5000)), though you're welcome to change that value to close the lock sooner or later, depending on your response time needs.

As before, refer to the code in the DoorLockServer.zip file. Open the project in Eclipse and focus on the AndroidDoorLockServerActivity class. Note the use of the try block that activates power to the PowerSwitch Tail for five seconds and makes the camerasurface.startTakePicture() call to the photo capture routine that will use the built-in Android camera.

AndroidDoorLock/DoorLockServer/src/com/mysampleapp/androiddoorlockserver/AndroidDoorLock¬ServerActivity.java
```
@Override
protected void loop() throws ConnectionLostException {

  if (mToggleButton.isChecked()) {
    if (LockStatus.getInstance().getLockStatus()) {
        try {
          powertail_.write(false);
              // pause for 5 seconds to keep the lock open
              sleep(5000);
              powertail_.write(true);
              LockStatus.getInstance().setMyVar(false);
              // Take a picture and send it as an email attachment
              camerasurface.startTakePicture();
              } catch (InterruptedException e) {
                    }
              }else {
                try {
                  sleep(10);
                } catch (InterruptedException e) {
                }
              }
        } else {
              powertail_.write(true);
          }
}
```

Compile and run this DoorLockServer project on your Android device. Start the web server on your Android device. Make sure it is properly connected to the IOIO board and the board is correctly wired to the PowerSwitch Tail. Access the IP address of the web server using any web browser that can access your local area network. If all goes according to plan, your electric door lock should unlock for several seconds and then relock. Cool!

We're two-thirds finished with this project. The final component is to take advantage of the fact that most Android devices (at least the Android phones) have a built-in camera. We're going to take advantage of that hardware asset by snapping a photo inside the door area several seconds after an unlock request and sending that photo to a designated email recipient. This way you know not only when an unlock request occurred but also who entered the door at the designated time.

Taking a Picture

For this part of the project, examine the CameraSurface.java file in the unzipped DoorLockServer directory. The key functions used to establish a camera surface and image capture are well documented in the Android SDK, and literally hundreds of Android photo-capturing code snippets and tutorials are available on the Internet.[18] I based the image capture portion of the web server application off of Android developer Krishnaraj Varma's Camera sample.

Setting up the camera for use in an Android application requires us to import several Android namespaces. To do so, we will need to perform a few additional steps to set up the display surface. The key libraries being used by the image capture portion of the program included in the DoorLockServer.zip file are as follows:

AndroidDoorLock/DoorLockServer/src/com/mysampleapp/androiddoorlockserver/CameraSurface.java
```
import android.content.Context;
import android.hardware.Camera;
import android.hardware.Camera.AutoFocusCallback;
import android.hardware.Camera.PictureCallback;
import android.hardware.Camera.ShutterCallback;
import android.util.AttributeSet;
import android.view.GestureDetector;
import android.view.MotionEvent;
import android.view.SurfaceHolder;
import android.view.SurfaceView;
import android.view.GestureDetector.OnGestureListener;
```

In addition to accessing the camera hardware itself, we also need to have the phone display a preview of the image being captured by the camera. To do so, we will first have to initialize the camera frame and surface variables:

```
private FrameLayout cameraholder = null;
private CameraSurface camerasurface = null;
```

---

18. http://developer.android.com/reference/android/hardware/Camera.html

These are used to allocate the surface and frame objects accordingly:

```
camerasurface = new CameraSurface(this);
cameraholder.addView(camerasurface, new
LayoutParams(LayoutParams.FILL_PARENT, LayoutParams.FILL_PARENT));
```

Krishnaraj uses callbacks to wait for certain operations to finish before proceeding. Examples of this include waiting for autofocus to set, waiting for the shutter to close, and waiting for the validation that image data has been successfully written to the SD card. The use of callbacks ensures that these events happen in serial fashion such that one won't begin until the other ends.

AndroidDoorLock/DoorLockServer/src/com/mysampleapp/androiddoorlockserver/CameraSurface.java
```java
public void startTakePicture(){
  camera.autoFocus(new AutoFocusCallback() {
    @Override
    public void onAutoFocus(boolean success, Camera camera) {
      takePicture();
    }
  });
}

public void takePicture() {
  camera.takePicture(
      new ShutterCallback() {
        @Override
        public void onShutter(){
          if(null != callback) callback.onShutter();
        }
      },
      new PictureCallback() {
        @Override
        public void onPictureTaken(byte[] data, Camera camera){
          if(null != callback) callback.onRawPictureTaken(data, camera);
        }
      },
      new PictureCallback() {
        @Override
        public void onPictureTaken(byte[] data, Camera camera){
          if(null != callback) callback.onJpegPictureTaken(data, camera);
        }
      });
  }
```

The act of writing data to the SD card occurs in the onJpegPictureTaken event. Since this image file is going to be sent as an email attachment and it's not necessary to store successive captures on the SD card, the image data is saved with the same filename each time a photo is taken.

```
FileOutputStream outStream = new FileOutputStream(String.format(
"/sdcard/capture.jpg"));

outStream.write(data);
outStream.close();
```

Note that if you prefer to store each progressive image capture on the phone's SD card rather than overwrite it with the same filename, you can append a timestamp to the suffix of the filename using Krishnaraj's original Camera code:

```
FileOutputStream outStream = new FileOutputStream(String.format(
"/sdcard/%d.jpg", System.currentTimeMillis()));
```

However, I don't recommend this approach unless you have plenty of storage capacity on your SD card and don't mind the duplication of image data on the phone and in your email inbox. If you do opt for this file naming method, you will also need to save the timestamped filename so you can later pass it when calling the email attachment instruction in the program. Now let's look at how to attach the image data to the email and send it.

Sending a Message

Now that we have captured and stored the camera-captured temporary image on the Android's SD card, we need to rely on a self-contained email routine that will email the attached photo without any user interface interaction. Fortunately for this project, we can call upon Jon Simon's JavaMail for Android-enhanced email routine. Download and reference the custom JavaMail for Android jar dependencies for Jon's email code to work properly.[19] We can then modify the code to account for our image attachment needs. To do so, we first need to import a number of Java libraries used by the JavaMail class:

```
import java.util.Date;
import java.util.Properties;
import javax.activation.CommandMap;
import javax.activation.DataHandler;
import javax.activation.DataSource;
import javax.activation.FileDataSource;
import javax.activation.MailcapCommandMap;
import javax.mail.BodyPart;
import javax.mail.Multipart;
import javax.mail.PasswordAuthentication;
import javax.mail.Session;
import javax.mail.Transport;
import javax.mail.internet.InternetAddress;
```

---

19. http://code.google.com/p/javamail-android/

```
import javax.mail.internet.MimeBodyPart;
import javax.mail.internet.MimeMessage;
import javax.mail.internet.MimeMultipart;
```

Methods for public Mail(String user, String pass) and public void addAttachment(String filename) throws Exception allow us to easily send the captured image file to a designated recipient. Sending a message is straightforward once the username, password, recipient, and attachment parameters are defined in onJpegPictureTaken() found in the AndroidDoorLockServerActivity.java file:

```
try {
    GMailSender mail = new GMailSender("YOUR_GMAIL_ADDRESS@gmail.com",
                    "YOUR_GMAIL_PASSWORD");
    mail.addAttachment(Environment.getExternalStorageDirectory() +
      "/capture.jpg");
    String[] toArr = {"EMAIL_RECIPIENT_ADDRESS@gmail.com"};
    mail.setTo(toArr);
    mail.setFrom("YOUR_GMAIL_ADDRESS@gmail.com");
    mail.setSubject("Image capture");
    mail.setBody("Image captured - see attachment");
    if(mail.send()) {
        Toast.makeText(AndroidDoorLockServerActivity.this,
                        "Email was sent successfully.",
                        Toast.LENGTH_LONG).show();
    } else {
        Toast.makeText(AndroidDoorLockServerActivity.this,
                        "Email was not sent.",
                        Toast.LENGTH_LONG).show();
    }
} catch (Exception e) {
    Log.e("SendMail", e.getMessage(), e);
}
```

Replace YOUR_GMAIL_ADDRESS@gmail.com, YOUR_GMAIL_PASSWORD, and EMAIL_RECIPIENT_ADDRESS@gmail.com with your Gmail account credentials. Note that the recipient does not have to be a Gmail user, so you can send the message to a non-Gmail account if you prefer to do so.

There are a few other preparatory instructions that are part of the email transmission process. Examine the downloaded code for a better understanding of all the dependencies and processes that take place to send a message from an Android device without user intervention.

Setting Hardware Permissions

We're almost done. By combining four separate Android programs into one, we are able to listen for an inbound HTTP request, unlock the electric door latch via the IOIO board, take a picture using the built-in camera on the Android device, and send that image as an email attachment.

With the photo capturing and email transmitting code in place, all that remains is to allow the program to access the camera, Wi-Fi radio hardware, and network to complete its task. As such, the AndroidManifest.xml file will need to contain permissions to access not only the network and Wi-Fi stack but also the camera and SD card:

```xml
<?xml version="1.0" encoding="utf-8"?>
<manifest xmlns:android="http://schemas.android.com/apk/res/android"
   package="com.mysampleapp.androiddoorlockserver"
   android:versionCode="1"
   android:versionName="1.0">
  <uses-sdk android:minSdkVersion="3" />
  <uses-permission android:name="android.permission."></uses-permission>
  <uses-permission android:name="android.permission.ACCESS_WIFI_STATE">
  </uses-permission>
  <uses-permission android:name="android.permission.INTERNET">
  </uses-permission>
  <uses-permission android:name="android.permission.WAKE_LOCK" />
  <uses-feature android:name="android.hardware.camera" />
  <uses-feature android:name="android.hardware.camera.autofocus"/>
  <uses-permission android:name="android.permission.CAMERA"/>
  <uses-permission android:name="android.permission.VIBRATE"/>
  <uses-permission
        android:name="android.permission.WRITE_EXTERNAL_STORAGE" />
  <application android:icon="@drawable/icon"
               android:label="@string/app_name">
    <activity android:name=".AndroidDoorLockServerActivity"
              android:label="@string/app_name"
              android:screenOrientation="landscape">
      <intent-filter>
        <action android:name="android.intent.action.MAIN" />
        <category
              android:name="android.intent.category.LAUNCHER" />
      </intent-filter>
    </activity>
  </application>
</manifest>
```

After setting the email account username, password, and recipient values as well as the IP address for your network, you can compile, install, and run the Android Door Lock server application on your Android smartphone.

Testing the Server

Test out the Android door lock server by accessing its base URL from a web browser. Verify that the electric lock releases and that the camera takes a photo and sends the image to the designated email recipient. If everything worked as expected, congratulate yourself on a job well done. Considering how many dependencies are involved with this project, getting everything to work just right the first time out is indeed a cause for celebration. If something went awry, carefully troubleshoot each function separately. Does the web server respond to requests? Does the PowerSwitch Tail electrify? Does the camera shutter snap? Also, depending on your network connection and the speed of Wi-Fi connectivity of your Android phone, it can sometimes take up to a minute to transmit the photo via email.

We have accomplished quite a bit of this project already, and for the most part, we could simply set a bookmark for the door lock URL and call it a day. But let's invest just a little more effort by creating a custom client for accessing the door lock URL like we did for the *Web-Enabled Light Switch* project. That way, we can quickly access the door lock via a one-click button. Indeed, we can begin amassing our home automation features into a über-controller mobile program that accesses our projects in a single collective interface.

9.5 Writing the Android Client

Writing code for the Android client to send unlock commands to the Android server is easy. Let's reuse code from the Web Enabled Light Switch Android client application to provide easy user access to the door latch function. This time, we will use a button instead of a toggle switch since we already programmed the lock to unlock for five seconds. This makes the toggle unnecessary. Another feature we will add to this application is to turn on the Wi-Fi radio if it isn't already active.

The basic flow of the program will be to launch it and check for Wi-Fi access. If the Wi-Fi radio is turned off, turn it on and wait for the client to connect to the network. Allow the user to press the displayed Unlock Door button, which will access the Android Door Lock server URL and unlock the door. Briefly, here are the steps we will take to code the unlock client:

1. Create an Android project in Eclipse called DoorLockClient.

2. Check if the Wi-Fi radio is on in the program's main activity. If Wi-Fi is turned off, activate it.

3. Add a Button to the main.xml layout description and label it "Unlock Door."

4. Reference the button in the DoorLock class and a listener for the button press event. If the Wi-Fi radio is being turned on for the first time when

> **Security Implications**
>
> One advantage of creating a custom Android client for unlocking a door is so that we also maintain (albeit very weak) security via obscurity to access the door lock. By not allowing a display of the bookmark URL on the screen when we access the web server, we keep its address hidden from nontechnical onlookers. However, this will only be adequate in low network security scenarios, since the URL itself is sent to the server in the clear. Later, in the Next Steps section of this chapter, one of the recommended enhancements is to consider adding better security to the project. The addition of a passcode or, better still, a sophisticated multifactor authentication scheme, will be a much better door lock system in the long run.

the program starts, keep the Unlock Door button disabled for a few seconds to allow the Wi-Fi interface to authenticate with the wireless access point and establish the client's IP address.

5. Add the URL request call in the button press event to the Android Door Lock server (ex: 192.168.1.230).

We will start by following the same procedure used in Section 7.6, *Writing the Code for the Android Client*, on page 115. Create a new Android project in Eclipse using the parameters shown in Figure 42, *Settings for the new Door Lock Client application*, on page 165.

Add a button called unlockbutton, label its text "Unlock Door," and set the button's width to fill the LinearLayout of the parent container. The main.xml file should look like this:

```xml
<?xml version="1.0" encoding="utf-8"?>
<LinearLayout xmlns:android="http://schemas.android.com/apk/res/android"
    android:orientation="vertical"
    android:layout_width="fill_parent" android:layout_height="fill_parent">
    <Button android:id="@+id/unlockbutton" android:layout_height="wrap_content"
        android:text="Unlock Door" android:layout_width="fill_parent"></Button>
</LinearLayout>
```

Save the changes. Open the DoorLockClient.java file and add references for the unlockbutton Button element and its event listener. Also add Wi-Fi radio detection and activation. The full listing for the DoorLockClient.java class should look like this:

AndroidDoorLock/DoorLockClient/src/com/mysampleapp/doorlockclient/DoorLockClient.java
```
package com.mysampleapp.doorlockclient;

① import java.io.InputStream;
   import java.net.URL;
   import android.net.wifi.WifiManager;
```

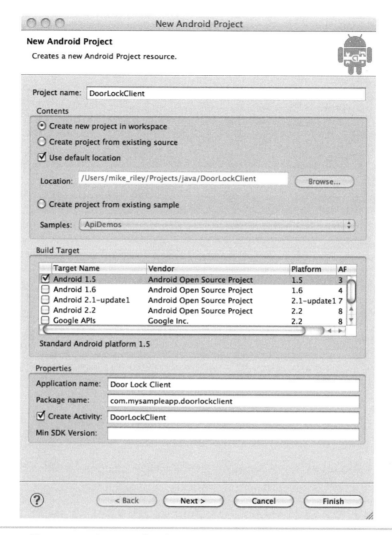

Figure 42—Settings for the new Door Lock Client application

```
import android.widget.Button;
import android.app.Activity;
import android.os.Bundle;
import android.util.Log;
import android.view.View;

public class DoorLockClient extends Activity {
    /** Called when the activity is first created. */
    @Override
    public void onCreate(Bundle savedInstanceState) {
        super.onCreate(savedInstanceState);
        setContentView(R.layout.main);
```

```
            Button unlockbutton = (Button) findViewById(R.id.unlockbutton);
            findViewById(R.id.unlockbutton).setOnClickListener
            (mClickListenerUnlockButton);
                try {
                    WifiManager wm =
                    (WifiManager) getSystemService(WIFI_SERVICE);
                    if (!wm.isWifiEnabled()) {
                        unlockbutton.setEnabled(false);
                        wm.setWifiEnabled(true);
                        // Wait 17 seconds for Wi-Fi to turn on and connect
                        Thread.sleep(17000);
                        unlockbutton.setEnabled(true);
                    }
                } catch (Exception e) {
                    Log.e("LightSwitchClient", "Error: " + e.getMessage(), e);
                }


        }
        View.OnClickListener mClickListenerUnlockButton =
            new View.OnClickListener() {
            public void onClick(View v) {
                try {
                    final InputStream is =
                    new URL("http://192.168.1.230:8000").openStream();
                }
                catch (Exception e) {
                }
            }
        };
}
```

① Import the library references for java.io.InputStream, java.net.URL and Android-specific android.widget.Button and android.net.wifi.WifiManager.

② Add a reference for the unlockbutton Button and assign it to the mClickListenerUnlockButton View method.

③ Query the state of the Wi-Fi radio, and if it's not active, turn the Wi-Fi radio on. Keep the unlockbutton disabled for seventeen seconds to allow enough time for the network connection to initialize.

④ Create the View.OnClickListener for the unlockbutton.

⑤ Request the Android door lock server address when the unlockbutton is clicked.

We have one more task to complete before we can test out the application. Remember how we had to set permission to access the network for the Web Enabled Light Switch Android client? We have to do the same thing for this Door Lock Client. We also have to grant permission to access the state of the

Wi-Fi radio as well. These permissions are noted in the AndroidManifest.xml file, which should look like this once these two permissions are added:

```xml
<?xml version="1.0" encoding="utf-8"?>
<manifest xmlns:android="http://schemas.android.com/apk/res/android"
  package="com.mysampleapp.doorlockclient"
  android:versionCode="1"
  android:versionName="1.0">
  <uses-permission android:name="android.permission.INTERNET" />
  <uses-permission android:name="android.permission.ACCESS_WIFI_STATE" />
  <uses-permission android:name="android.permission.CHANGE_WIFI_STATE" />

  <application android:icon="@drawable/icon" android:label="@string/app_name">
    <activity android:name=".DoorLockClient"
            android:label="@string/app_name">
      <intent-filter>
            <action android:name="android.intent.action.MAIN" />
            <category android:name="android.intent.category.LAUNCHER" />
      </intent-filter>
    </activity>
  </application>
</manifest>
```

Save the project and test it using an available Android phone. First test its operation with the Wi-Fi radio turned on. The button should be instantly accessible after the program has launched. Quit the program, preferably using a task manager (i.e., make sure the running instance of the program is destroyed and not running silently in the background). Next, turn off the Wi-Fi radio and launch the Door Lock Client application again. This time, the Unlock Door button will be disabled while the program turns on the Wi-Fi radio and waits until a connection with the network has been established. If the radio turned on, you're ready for a live test with the door lock server.

Click the Unlock Door button. Within a second or two, the electric lock should click open and then close approximately five seconds later. If it did, congratulations on a job well done! If it didn't, verify that your Android device is indeed connected to the network. Test the URL access via the Android web browser. If you can't access the URL, make sure the Android door lock server is still set to the static IP we defined earlier and that it is running. Try accessing the URL from a different system just to verify that the rest of your home network can access the Android door lock server.

9.6 Test and Install

Now that the Android client is transmitting the URL request via the toggle button interface, we are nearly finished with this project. Test the lock and photo capture mechanism by powering up the Android phone server, making sure all the connections between the phone, IOIO board, and PowerSwitch Tail are connected. Send a request to the Android phone server from the other Android device running the Android Door Lock client. Note that in order to test this request successfully, the client Android device needs to be connected to the same Wi-Fi access point as the Android phone server. Your test rig may look like mine (Figure 43, *Testing the Android Door Lock*, on page 169).

The final task is to actually install the lock mechanism in the doorframe. This can be a daunting process if you are not comfortable with boring out wood and properly routing electrical wiring behind drywall. In fact, I strongly recommend that if you intend to permanently install this hardware configuration, contact a reputable carpenter and electrician to assist with the installation. The extra money you spend will be well worth the safety and security of your home, and it will keep your sanity intact.

When installing the lock, keep the Android phone, IOIO board, and PowerSwitch Tail in an easy, accessible location. It should go without saying that you shouldn't place these in the wall in case you need to service or replace any of these components in the future, not to mention that they could pose a fire hazard if the wiring is not correctly shielded. I suggest obtaining a project box from an electronics supplier. The box should be large enough to fit all the components, with room for expansion should you need to house additional hardware for project enhancements. Always practice safe wiring techniques. Once a circuit is well established, I prefer soldering components in place and then covering the exposed conductive surface, like circuit leads and stripped wiring, with heat shrink tubing to prevent any shorting of the circuit. If you have an electrician assist with the hardware installation, consult with this professional about best practices and recommendations as well.

9.7 Next Steps

Congratulations! You have just completed one of the most complex projects in this book. You have come a long way and acquired a great deal of knowledge and experience. You now have the ability to automate a variety of electrical devices in your own home. Our final project will combine a number of these techniques to create an application that will listen for a number of events and relay these to you via a text-to-speech interface. But before we get started, consider expanding your Android Door Lock with these additional features:

Figure 43—Testing the Android Door Lock

- Implement Steve Gibson's Perfect Paper Passwords to provide a more secure, multifactor, one-time password authentication scheme.[20] By using the Perfect Paper Password approach, you will be able to share one-time use entry codes to anyone requiring secure access to your home, such as visiting health professionals, house cleaning service personnel, and maintenance workers.

- Connect a PIR sensor to the IOIO board to capture and transmit motion-detected events. While the current design does something similar, it can be problematic if entry to the target area is intentionally or unintentionally delayed. Take advantage of the numerous other analog and digital pins on the IOIO board and hook up a PIR sensor like the one we used in Chapter 4, *Electric Guard Dog*, on page 43.

- Attach more than one electric door lock to the IOIO Web Server program and access these locks via different URL paths. For example, open the front door by accessing http://192.168.1.230/frontdoor, and the cellar door via http://192.168.1.230/cellardoor.

- Go beyond just controlling door locks from an Android phone. Electrify lights, appliances, computers, and any other electrical device in your home via the IOIO web server. Expand the web server program on your Android phone to log events, email status updates, or detect orientation changes (i.e, someone or something moved the phone) via Android's compass and accelerometer sensors.

---

20. https://www.grc.com/ppp.htm

CHAPTER 10

Giving Your Home a Voice

Wouldn't it be cool to walk into your front door and be greeted by your home, having it inform you of any important events that occurred while you were away? How about asking your home to check your email inbox status, read you the weather forecast, or queue up your favorite music on the stereo? It could also audibly inform you of triggered sensors in real time, such as telling you about water-level alerts or birdseed refillings from our first two projects. (See Figure 44, *Event notification*, on page 172.)

This project will bring those fanciful ideas to life. We will create a central hub capable of relaying the communication from all the other projects we built in this book and do so in a natural speaking voice.

Receiving emails and tweets about what's going on in your home is pretty neat, but wouldn't it be even cooler if you could have a conversation with your home? What if you could ask it questions like "What time is it?" or dictate commands like "Turn on the lights" or "Listen to music" and have your home respond in kind. That's what we're going to program in this capstone project that brings together network-enabled projects like the Web-Enabled Light Switch and the Android Door Lock and controls them via voice command.

While we're at it, we will hook into a few other nice-to-have vocal commands like selecting musical artists and their respective albums for audio playback on the stereo, turning up and down the volume, and so on.

10.1 What You Need

While we could develop this project on Windows using Microsoft's Speech API or on Linux using the open source Festival project, I chose the Mac platform because I personally find the Text-to-Speech (TTS) renditions in OS X 10.7 (aka Lion) to be the best of the voices that ship between the three operating

Figure 44—Event notification. Let your home tell you when automation events occur with your projects.

systems. Most Mac users don't know these voices exist, let alone that downloading additional OS X voices can expand your choices.

Here are the items you will need to put this project into action:

- An Apple Mac computer running OS X 10.7 (Lion) or higher
- A home stereo with standard 3.5mm or RCA audio input jacks
- One of the following:
    - A male-to-male stereo miniplug cable to connect the Mac to your home stereo, or
    - A 3.5mm stereo headphone-to-RCA adapter cable if your stereo only uses standard RCA input jacks, or
    - A wireless Bluetooth speaker, such as the Supertooth DISCO[1]
- A wireless microphone and receiving station, such as the Radio Shack Wireless Lapel Microphone System[2]
- A 3.5mm headphone-to-USB adapter to send the wireless mic station's audio into the Mac, such as Griffin Technology's iMic[3]

---

1. http://www.supertooth.net/AU/produitmusique.htm
2. http://www.radioshack.com/product/index.jsp?productId=2131022
3. http://store.griffintechnology.com/imic

Before we listen to the computer voices, we first need to be able to reproduce the Mac's spoken audio on a set of speakers, whether they be attached to the Mac, connected via a stereo, or transmitted to a wireless Bluetooth speaker.

10.2 Speaker Setup

The speaker being used to amplify the computer's audio is a key factor in this project's success. The speaker needs to be loud enough to be heard from one or more rooms away and ideally should be heard throughout the house if possible. Let's take a look at both wired and wireless approaches.

The quickest way to connect a Mac computer up to a home stereo is by using a male-to-male 3.5mm stereo headphone to an RCA adapter cable. This cable will run from the Mac's headphone jack to the stereo amplifier and/or receiver's audio input jack. If your home stereo doesn't support a 3.5mm input jack, you will need a 3.5mm female-to-RCA-male audio cable. The length of the cable needs to comfortably run from the computer to the stereo, so take that into consideration when purchasing the cable from your preferred audio-video supplier.

If the computer and stereo are separated by several rooms, you will be running a lot of wire and probably need to drill a few holes in the process. If you don't want to operate the computer in the same room as the stereo and don't like the idea of fishing wiring through walls to connect the two, consider a wire-free alternative. If this is your situation, I recommend using the external Bluetooth speaker option since it offers the most flexibility.

Pairing an external Bluetooth speaker with the Mac is easy. Simply turn on Bluetooth on the Mac via the Bluetooth System Preference pane. Then power up the external Bluetooth speaker and set it to pair with your computer. This is typically done by holding down the power button on the speaker until the speaker's Bluetooth indicator light starts flashing. Then click the Set Up New Device... button on the Mac's Bluetooth Preference Pane. This should auto-detect the Bluetooth speaker. In the case of the Supertooth DISCO speaker, it displays "ST DISCO R58" on my Mac, as shown in Figure 45, *Bluetooth wireless speaker pairing*, on page 174.

Select the speaker name. Depending on the Bluetooth speaker you are connecting to, it may automatically establish a connection or it may require a four-digit confirmation code such as 0000 or 1234 to be typed in on the screen. In the case of the Supertooth DISCO speaker, my Mac automatically configured the speaker without requiring any confirmation codes.

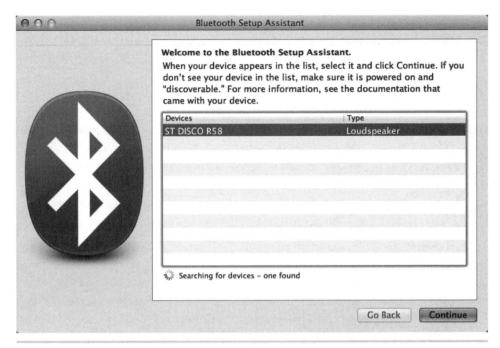

Figure 45—Bluetooth wireless speaker pairing

If the discovery and configuration went smoothly, you should receive a confirmation message on the screen that pairing with the speaker was successful. Go to the Sound option in System Preferences and select the speaker in the Output tab. Then use the iTunes music application on your Mac to play back audio and verify that you can indeed hear the sound reproduced on the paired speaker. Conclude your pairing confirmation testing by entering the Speech option in System Preferences and then selecting the Text to Speech tab. Click the Play button. If you hear the TTS playback on the speaker, your talking Mac hardware setup is successfully configured. Set the volume on the speaker and dial up or down the output volume level on the Mac to get the sound output just right for the audio coverage area you have in mind.

There are trade-offs between these wired and wireless audio configurations. If you need the convenience of a wire-free audio transmission, the external Bluetooth speaker option is the way to go. But if you value high fidelity sound over wireless convenience, a wired connection to a dedicated stereo amplifier/receiver offers the best sound reproduction to multiple speaker outputs. If you are fortunate enough to have already prewired the rooms in your home for stereo sound, the wired computer-to-stereo approach is the obvious choice.

> **Why Bluetooth Audio?**
>
> Bluetooth wireless audio capabilities are available in most Mac computers. Using it will give you maximum flexibility when placing the computer and external Bluetooth speaker in different locations. Rather than running an audio cable from the computer to a stereo, we can rely on high fidelity wireless Bluetooth audio that can broadcast up to thirty feet away.

Next we'll configure the Mac to listen for voice commands and respond with a high-quality voice response. Then we will write an AppleScript script that will leverage OS X's built-in speech recognition server to listen for specific commands and act on them accordingly.

10.3 Giving Lion a Voice

Before we can talk to a Mac, we must first enable its speech recognition server. Note that the speech recognition server had been broken on the OS X 10.5 and 10.6 releases and was finally fixed in the 10.7 Lion release. This once again makes the Mac a viable speech recognition platform.

In order to configure the Mac to use its speech recognition capabilities, click on the Speech icon in the System Preferences panel, as shown in Figure 46, *Accessing OS X speech settings*, on page 176.

Select the Speech Recognition tab and turn on the Speakable Items, as shown in Figure 47, *Turn on speakable items*, on page 177.

Read the tips dialog box that is displayed the first time you enable this option and take heed of the recommendations. Speech recognition algorithms are not yet powerful enough to effortlessly understand a variety of dialects, accents, and volume levels, but the technology is getting better all the time.

I find I have to be especially loud and clear when speaking to the Mac, making an effort to slowly enunciate my commands with almost no background noise. You may also need to play around with microphone gain and placement from your mouth as well as acclimate to the cadence for the speech recognizer to work with the vocabulary we'll be defining in our script. Note that the Mac sets the Microphone setting to use the Internal Microphone by default. We will revisit this setting later when we change this to use the iMic adapter, but it's okay to leave the setting as it is for now.

When the Speakable Items option is activated, you will see a round microphone graphic appear on your computer's screen. This is the speech recognizer

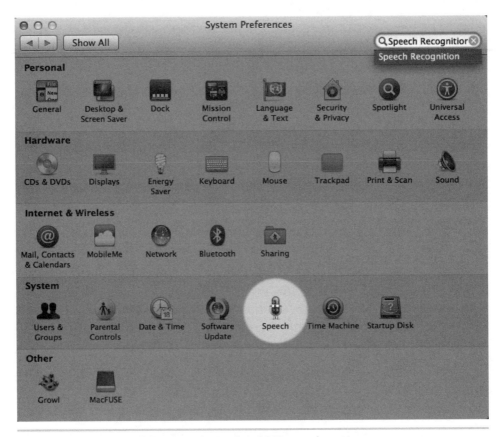

Figure 46—Accessing OS X speech settings

window. You activate the recognizer by holding down the Escape key on the keyboard. We will remove this keyboard requirement once we have our talking home script running and our wireless microphone set up, but for now we'll leave it be so that we can more easily debug our script.

Before closing the Speech preference panel, we have one more option to set. Click on the Text to Speech tab (see Figure 48, *Text to Speech settings*, on page 178) and select a System Voice from the drop-down list.

You can preview voices by clicking the Play button. The default voice is Alex. It's pretty good, but I prefer the American female voice Samantha. Since the voice files are quite large, Apple doesn't ship all of the selections with Lion. Instead, you have to obtain them by selecting the Customize… System Voice menu option. Doing so will display the dialog box shown in Figure 49, *Lion voice selections*, on page 179, which lists the voices freely available for download from Apple.

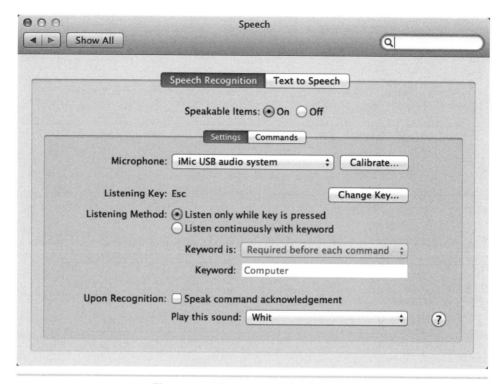

Figure 47—Turn on speakable items.

There are plenty of voices to choose from, and you can preview each one before downloading them by clicking the Play button. Once you select a voice that you like, it may take a while to download and configure the selected voice files on your computer, depending on your Internet connection and Mac CPU speed. As an example, the Samantha voice file is over 450 megabytes in size.

Once downloaded and installed, you can further tweak the voice playback by moving the Speaking Rate slider for faster or slower playback. I suggest keeping it on the normal default for now and modifying it if necessary once you have the whole wireless mic rig and speaker system working. Speaking of which, our next task is to get the wireless mic hooked up and calibrated for speech recognition.

10.4 Wireless Mic Calibration

If you're using a MacBook Pro or iMac computer, you could use the computer's internal microphone. It works okay if you are sitting directly in front of the laptop, but it falters the farther you are from the screen. We need the mobility

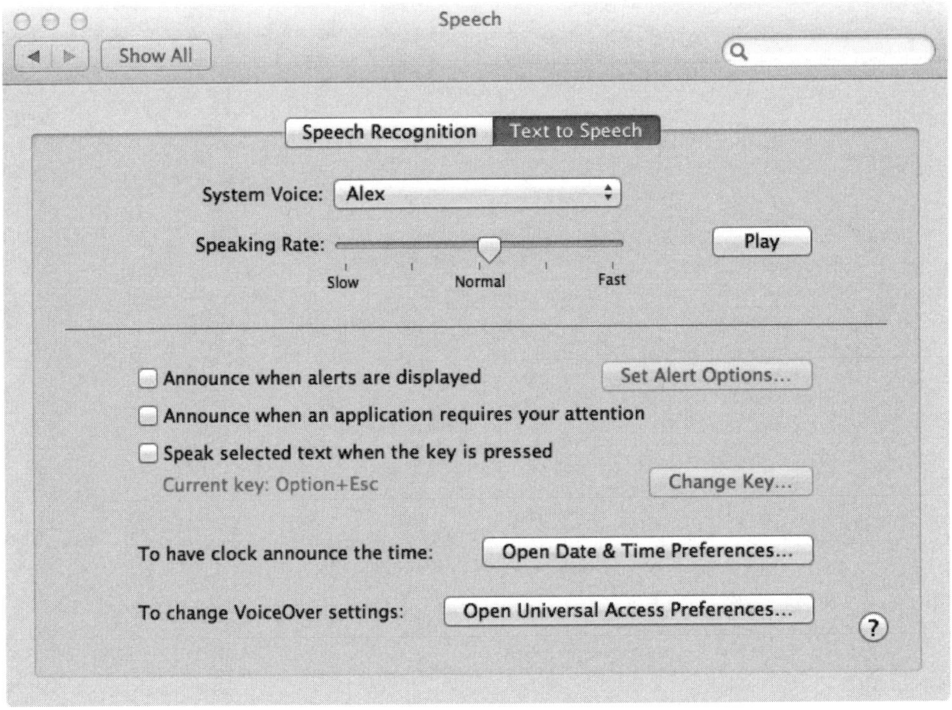

Figure 48—Text to Speech settings

of being able to converse with our home while walking around, watching TV, making breakfast in the kitchen, or cleaning the living room. This will be accomplished by using a wireless microphone.

For a wireless microphone to reliably work with the Mac's speech recognizer, it needs to be a decent quality wireless mic with clear audio signal transmission. A mic that delivers crackling, hissing audio won't work too well because the speech recognizer will struggle to distinguish between the signal and the noise. If this is a project you're planning on fruitfully using for a long time, invest in a quality wireless microphone, like those used by professional singers. These can cost over two hundred dollars or more, depending on the features and broadcast range, but they make a big difference in consistent delivery of clear audio. For those interested in testing the waters before committing that kind of money in audio hardware, the Radio Shack Wireless Lapel Microphone System is a more economical compromise.

Plug the Griffin iMic adapter into one of the Mac computer's available USB ports, then plug the output of the wireless base station into the iMic's miniplug inputs. Make sure that the mic input is selected on the iMic, turn on the

Figure 49—Lion voice selections

wireless mic and power up the base station. Select the iMic USB audio system in the Speech Recognition System Preferences pane and click the Calibrate... button. This will bring up the microphone calibration window (Figure 50, *Microphone calibration*, on page 180). Speak into the microphone while moving the slider left or right to keep the audio level in the green bar area.

Try moving around the room wearing the active wireless mic and verify that the calibration bars still stay within the green while speaking.

We have one more task to complete before we can start coding. Let's hook up the speakers for audio output.

10.5 Programming a Talking Lion

Writing good speech synthesis and recognition software is hard. That's why we're going to take advantage of all the hard work Apple speech software engineers have poured into OS X. The engine can be accessed a variety of ways, via the preferred method of Objective-C to Perl, Python, Ruby, and other scripting language hooks. But the easiest way I have found to tinker and quickly modify and test on the fly is via AppleScript.

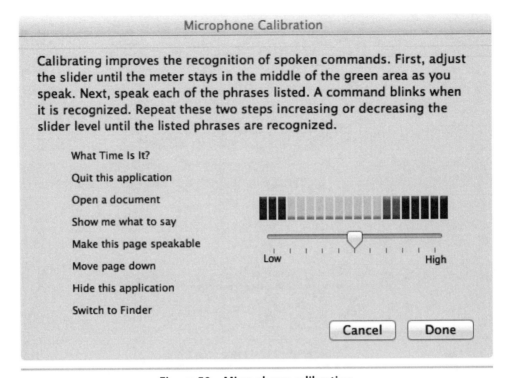

Figure 50—Microphone calibration

I'll be the first to admit that I am not a big fan of AppleScript. Its attempt to turn script writing into a natural English sentence structure works only on a superficial level. It breaks down pretty quickly for any intermediate developer fluent in more elegant scripting languages like Ruby or Python. Even simple tasks like string manipulation turn out to be a real pain in AppleScript. That said, AppleScript trumps these other languages when it comes to effortless automation integration with other AppleScript-aware OS X applications. Bundled programs like iTunes, Mail, Safari, and Finder are fully scriptable, as are a number of third-party OS X programs like Skype, Microsoft Office, and the like. In the case of this project, Apple's speech recognition server is also highly scriptable, and that's what we're going to call upon in this project to make the magic work.

While AppleScript can be written using any text editor, it should come as no surprise that it's best hosted within the AppleScript Editor application. This can be found in the Applications/Utilities folder. Launching the AppleScript editor for the first time will open a blank, two-pane coding window. The top half of the window is used to enter code, while the bottom consists of three tabs for monitoring events, replies, and results of the executing script. The editor aids in writing script by color coding AppleScript syntax, but it doesn't offer IDE-friendlier features like code completion or on-the-fly compiling. Fortunately, scripts are typically short, so these omissions are not crippling.

AppleScript has its own vocabulary, keywords, and idioms. Learning AppleScript isn't difficult, but it can get maddening at times when you have to massage the syntax just right to make the script do what you intended. For example, parsing a string for an email address is easy in most scripting languages. Not so in AppleScript. Partly due to its historical ties and partly due to the way AppleScript expects you to work, it's complicated. So with regard to the code we will write for this project, you will just have to trust me and try to follow along. If you find AppleScript to your liking or want to see what else it can do to further extend the code for this project, review Apple's online documentation for more information.[4]

Before writing the script, let's think about what we want it to do. First, we want it to respond to a select group of spoken words or phrases and act on those commands accordingly. What commands should we elicit? For starters, how about having the script hit the URLs we exposed in some of our networked projects, like the Web-Enabled Light Switch or the Android Door Lock? While we're at it, let's make use of some of the bundled OS X applications like Mail and iTunes to check and read our unread email and play music we want to hear. Let's also ask our house what time it is.

We need to initialize the SpeechRecognitionServer application and populate the set of words or phrases that we want it to listen to. Using a series of if/then statements, we can react to those recognized commands accordingly. For example, if we ask the computer to play music, we will call upon the iTunes application to take an inventory of music tracks in its library, sort these by artist and album, populate these as more words/phrases to interpret, and have the text-to-speech engine ask us which artist and album we want to listen to. Similarly, we can have our unread email read to us via a check mail command. Doing so will launch the Mail application, poll your preconfigured Mail accounts for new mail, check the inbox for unread messages, and perform a text-to-speech reading of unread sender names and message titles.

Now let's take a closer look at the details of the script's execution. Here's the full script in its entirety. Most of the syntax should be easy to follow, even if you are not familiar with AppleScript.

---

4. http://developer.apple.com/library/mac/#documentation/AppleScript/Conceptual/AppleScriptLangGuide/introduction/ASLR_intro.html

GivingYourHomeAVoice/osx-voice-automation.scpt

```applescript
with timeout of 2629743 seconds
  set exitApp to "no"
  repeat while exitApp is "no"
①    tell application "SpeechRecognitionServer"
      activate
      try
        set voiceResponse to listen for {"light on", "light off", ¬
          "unlock door", "play music", "pause music", ¬
          "unpause music", "stop music", "next track", ¬
          "raise volume", "lower volume", ¬
          "previous track", "check email", "time", "make a call", ¬
          "hang up", "quit app"} giving up after 2629743
      on error -- time out
          return
      end try
    end tell

②    if voiceResponse is "light on" then
      -- open URL to turn on Light Switch
      open location "http://192.168.1.100:3344/command/on"
      say "The light is now on."

    else if voiceResponse is "light off" then
      -- open URL to turn off Light Switch
      open location "http://192.168.1.100:3344/command/off"
      say "The light is now off."

    else if voiceResponse is "unlock door" then
      -- open URL to unlock Android Door Lock
      open location "http://192.168.1.230:8000"
      say "Unlocking the door."

③    else if voiceResponse is "play music" then
      tell application "iTunes"
        set musicList to {"Cancel"} as list
        set myList to (get artist of every track ¬
          of playlist 1) as list
        repeat with myItem in myList
          if musicList does not contain myItem then
            set musicList to musicList & myItem
          end if
        end repeat
      end tell

      say "Which artist would you like to listen to?"
      tell application "SpeechRecognitionServer"
        set theArtistListing to ¬
          (listen for musicList with prompt musicList)
      end tell
```

```
            if theArtistListing is not "Cancel" then
              say "Which of " & theArtistListing & ¬
                  "'s albums would you like to listen to?"
              tell application "iTunes"
                tell source "Library"
                  tell library playlist 1
                    set uniqueAlbumList to {}
                    set albumList to album of tracks ¬
                        where artist is equal to theArtistListing

                      repeat until albumList = {}
                        if uniqueAlbumList does not contain ¬
                          (first item of albumList) then
                          copy (first item of albumList) to end of ¬
                              uniqueAlbumList
                        end if
                        set albumList to rest of albumList
                      end repeat

                      set theUniqueAlbumList to {"Cancel"} & uniqueAlbumList
                      tell application "SpeechRecognitionServer"
                        set theAlbum to (listen for the theUniqueAlbumList ¬
                            with prompt theUniqueAlbumList)
                      end tell
                  end tell
                  if theAlbum is not "Cancel" then
                    if not ((name of playlists) contains "Current Album") then
                    set theAlbumPlaylist to ¬
                        make new playlist with properties {name:"Current Album"}
                    else
                      set theAlbumPlaylist to playlist "Current Album"
                      delete every track of theAlbumPlaylist
                    end if
                    tell library playlist 1 to duplicate ¬
                      (every track whose album is theAlbum) to theAlbumPlaylist
                      play theAlbumPlaylist
                  else
                    say "Canceling music selection"
                  end if
                end tell
              end tell
            else
              say "Canceling music selection"
            end if
④       else if voiceResponse is "pause music" or ¬
          voiceResponse is "unpause music" then
          tell application "iTunes"
            playpause
          end tell
```

```
      else if voiceResponse is "stop music" then
        tell application "iTunes"
          stop
        end tell

      else if voiceResponse is "next track" then
        tell application "iTunes"
          next track
        end tell

      else if voiceResponse is "previous track" then
        tell application "iTunes"
          previous track
        end tell

      -- Raise and lower volume routines courtesy of HexMonkey's post:
      -- http://forums.macrumors.com/showthread.php?t=144749
⑤     else if voiceResponse is "raise volume" then
        set currentVolume to output volume of (get volume settings)
        set scaledVolume to round (currentVolume / (100 / 16))
        set scaledVolume to scaledVolume + 1
        if (scaledVolume > 16) then
          set scaledVolume to 16
        end if
        set newVolume to round (scaledVolume / 16 * 100)
        set volume output volume newVolume
      else if voiceResponse is "lower volume" then
        set currentVolume to output volume of (get volume settings)
        set scaledVolume to round (currentVolume / (100 / 16))
        set scaledVolume to scaledVolume - 1
        if (scaledVolume < 0) then
          set scaledVolume to 0
        end if
        set newVolume to round (scaledVolume / 16 * 100)
        set volume output volume newVolume

⑥     else if voiceResponse is "check email" then
        tell application "Mail"
          activate
          check for new mail
          set unreadEmailCount to unread count in inbox
          if unreadEmailCount is equal to 0 then
            say "You have no unread messages in your Inbox."
          else if unreadEmailCount is equal to 1 then
            say "You have 1 unread message in your Inbox."
          else
            say "You have " & unreadEmailCount & ¬
              " unread messages in your Inbox."
          end if
```

```
            if unreadEmailCount is greater than 0 then
              say "Would you like me to read your unread email to you?"
              tell application "SpeechRecognitionServer"
                activate
                set voiceResponse to listen for {"yes", "no"} ¬
                    giving up after 1 * minutes
              end tell
              if voiceResponse is "yes" then
                set allMessages to every message in inbox
                repeat with aMessage in allMessages
                  if read status of aMessage is false then
                    set theSender to sender of aMessage
                    set {savedDelimiters, AppleScript's text item delimiters} ¬
                        to {AppleScript's text item delimiters, "<"}
                    set senderName to first text item of theSender
                    set AppleScript's text item delimiters ¬
                        to savedDelimiters
                    say "From " & senderName
                    say "Subject: " & subject of aMessage
                    delay 1
                  end if
                end repeat
              end if
            end if
          end tell

⑦        else if voiceResponse is "time" then
            set current_time to (time string of (current date))
            set {savedDelimiters, AppleScript's text item delimiters} to ¬
                {AppleScript's text item delimiters, ":"}
            set hours to first text item of current_time
            set minutes to the second text item of current_time
            set AMPM to third text item of current_time
            set AMPM to text 3 thru 5 of AMPM
            set AppleScript's text item delimiters to savedDelimiters
            say "The time is " & hours & " " & minutes & AMPM
⑧          --else if voiceResponse is "make a call" then
          --   tell application "Skype"
          --   -- A Skype API Security dialog will pop up first
          --   -- time accessing Skype with this script.
          --   -- Select "Allow this application to use Skype" for ¬
          --   -- uninterrupted Skype API access.
          --      activate
          --      -- replace echo123 Skype Call Testing Service ID with ¬
          --      -- phone number or your contact's Skype ID
          --      send command "CALL echo123" script name ¬
          --         "Place Skype Call"
          --   end tell
          -- else if voiceResponse is "hang up" then
          --   tell application "Skype"
```

```
        --        quit
        --    end tell
⑨    else if voiceResponse is "quit app" then
        set exitApp to "yes"
        say "Listening deactivated. Exiting application."
        delay 1
        do shell script "killall SpeechRecognitionServer"
    end if
  end repeat
end timeout
```

① The first thing we should do to keep the script running continuously is wrap the script in two loops. The first is a with timeout... end with loop to prevent the script from timing out. The timeout duration must be set in seconds. In this case, we're going to run the script for one month (there are roughly 2.6 million seconds in an average month).

The second loop is a while loop that repeats until the exitApp variable is set to yes via the "Quit app" voiceResponse, as shown toward the end of the code listing.

Next, initialize the Speech Recognizer Server and pass it an array of the key words and phrases via the listen for method. We will keep the recognizer alive for a month so it can await incoming commands without having to restart the script when the listening duration times out. You can extend this month-long duration by changing the giving up value.

② If the incoming phrase is interpreted as lights on, we will open the default browser and direct it to the on URL of our web-enabled light switch. "Lights off" will request the off URL from that project. We can also perform the same open location URL call for the Android door lock project too.

③ Besides triggering URL calls via voice, we can also interact with AppleScript-able OS X applications like iTunes and Mail. In this code snippet, we do the following:

1. Open iTunes.
2. Create an empty list array.
3. Populate that array with every song track in our local iTunes library, eliminating duplicate titles along the way.
4. Extract the artist names from the array of tracks.
5. As long as there is at least one artist in the array, pass the array of artist names to the speech recognition server via its listen for method.

6. Ask the user to pick an artist to listen to. If the user responds with the name of an artist in the library, populate the speech recognizer with the name(s) of that artist's album(s). Users can also exit the play music routine at this point by saying the word "Cancel."

7. If an artist has more than one album in the library, use the same type of procedure as the artist selection process to select the desired artist's album. Otherwise, start playback of the album immediately.

④ The pause/unpause and stop music commands, along with the next and previous track commands, call iTunes's similarly named methods.

⑤ The raise and lower volume commands capture the Mac's current output volume and raises or lowers it equivalent to a single press of the up and down volume keys on the Mac's keyboard. These commands are especially helpful when having to raise or lower music playback volume hands-free.

⑥ This portion of the script expects that you have already configured your desired email accounts to work with OS X's built-in Mail application. In the Mail snippet, we do this:

1. Open Mail.
2. Poll all configured mail servers for new, unread email messages.
3. Count the number of unread mail messages in the unified inbox and speak that amount.
4. If there are any unread messages, ask users if they would like to have their unread messages read to them.
5. If the user answer is yes, create an array of the unread messages and read the name and the subject line of the email. Otherwise, exit the routine.

⑦ This routine extracts the current time from AppleScript's current date routine. From there, we do this:

1. Assign the current time to the string current_time.
2. Use AppleScript's savedDelimiters function to split the current_time string via the : delimiter. This breaks the string apart into its constituent hour and minute values. The remainder of the string contains the a.m. or p.m. designation.
3. Assign these time values to their appropriate variables (hours, minutes, AMPM) and speak them accordingly.

⑧ Uncomment these lines (remove the double-dash [–] characters used to indicate a comment in AppleScript) if you have the Mac Skype client installed and you want to place a hands-free call. Configure the account name of your choice in the echo123 Skype call testing service account.

⑨ This command exits the script and ensures that the speech recognition server process is indeed killed by issuing a `killall SpeechRecognitionServer` command from the shell.

Once you have entered the script in the AppleScript editor, save it and click the Compile button on the editor's toolbar. If the script contains any typos or errors, it will fail to compile. Clean up whatever problems occur and make sure you attain a clean compile. Also make sure that your calibrated wireless headset is turned on and the input audio levels are properly set. Turn up the volume on your external speakers loud enough to hear the responses and music playback. Then click the Run button and get ready to talk.

10.6 Conversing with Your Home

The culminating moment of glory has arrived. Speak the command "Time," and listen for your computer to respond with the current time. Ask it to "play music," and your computer should respond with "Which artist would you like to listen to?" Respond in kind with an artist in the computer's iTunes library and select the album from which to start playing. Say "Stop music" when you're done listening to the music.

Query for unread email in your inbox by asking your computer to "check mail." Check the computer screen to verify that the computer responds with the correct count and reads the correct email sender and subject lines.

If you have your Android door lock or web-enabled light switch running, say "Unlock door" or "Light on" to watch your door unlock or light turn on accordingly. You now have your own voice-activated home. Pretty cool!

Try issuing commands from different locations. Move around the room, then try from other rooms. See how far your wireless microphone's signal will reach before it starts to cut out and your commands are no longer being acknowledged. Keep these boundaries in mind when interacting with your computer.

To give the script more permanence, convert it to an executable. Place its icon in the OS X desktop dock and control-click its icon to select Open at Login from the Options section of the pop-up menu. This will automatically launch the script each time you log into your Mac's desktop, ready to listen to your predetermined voice commands.

Continue to tweak the script and add any new phrases and functionality that best suit your environment. Network-enable the Curtain Automation project and instruct your home to "open drapes" or add a weather option that pulls down the weather forecast from the National Oceanic and Atmospheric Administration's weather.gov website and reads it aloud. Consider adding more features as suggested in the Next Steps section.

10.7 Next Steps

Kudos to you for completing the last project in the book. As you have seen, enabling voice recognition is a remarkably trivial matter and while the technology isn't perfect, it's still pretty wild that we have the ability to control our home in ways that were considered science fiction twenty years ago.

Take this higher-level automation skill further by pursuing the following enhancements:

- Expand the spoken email routine to include saying the timestamp and message body. Add the ability to delete a message or reply to an email with prepared templates (ex: "Reply yes").

- Duplicate the script's iTunes artist/album lookup array functionality for the Skype client so that you can place hands-free calling to anyone on your Skype contact list. After saying "Make a call," the script will populate a listen for array with the contact names in your active Skype account. Like the artist name response, reply with the name of the contact you want to call and the script will automate Skype to do so.

- Add Text to Speech extensions to the Tweeting Bird Feeder and Package Delivery Detector Python scripts that speak status updates when events like bird landings or package deliveries are detected. This can be done via an Open Script Architecture (OSA) shell command.[5]

- Bring speech recognition to other computing platforms besides Apple OS X by converting the script to an Android application by calling upon Android's RecognizerIntent intent or Microsoft's Speech API for the Windows platform.[6]

---

5. http://developer.apple.com/library/mac/#documentation/Darwin/Reference/ManPages/man1/osascript.1.html
6. http://developer.android.com/resources/articles/speech-input.html or http://msdn.microsoft.com/en-us/library/ee125663%28VS.85%29.aspx, respectively.

Part III

Predictions

CHAPTER 11

Future Designs

The majority of this book has focused on real home automation projects that you can build inexpensively today. This chapter takes a look at the exciting, rapid evolution of microcontrollers, smartphones, and computers and then forecasts what's on the horizon for these technologies.

We will take a look at the near-term prospects of Arduinos, Androids, and computer operating systems and then extrapolate these developments out roughly a decade to see how these products will help form the foundation of high-tech residential living in the year 2025. Looking back at how sophisticated mobile technology has become over the last ten years, the ideas in this future setting may not be as farfetched as they sound. In fact, most of the proposed scenarios could be implemented today, with the projects you learned about in this book helping to point the way. But first, let's take a look at what's coming up in the next year or so.

11.1 Living in the Near

The open hardware movement is rapidly gaining momentum, and more businesses and services are expanding or being created as a result. The established technologies like Arduino and Android are not standing still either. Both of these platforms had major version upgrades just as this book was finishing its production.

This section talks about what changes are in store with these near-term technology innovations and how these new releases will impact anyone choosing to use these as the technologies of choice when constructing the projects in this book.

Arduino 1.0

Just as the final pages in this book were written, the Arduino team announced the impending release of Arduino version 1.0. A number of substantial changes have been made in this version that will certainly create legacy code nightmares. This was a bold move by the Arduino team given the considerable amount of user-generated libraries, code samples, documentation, books, and videos made using earlier Arduino releases. As a consequence, this book is no different. Once Arduino 1.0 is widely adopted, the project and library dependencies will almost certainly need to be rewritten to support the changes. Most notable of these changes include the following:

- Sketch file name extensions have been changed from .pde to .ino. This was done to avoid confusion with Processing sketches that also use the same .pde extension.

- The Arduino Ethernet library will natively support DNS and DHCP. This will make IP address assignments vastly easier.

- The String class has been optimized so that it requires fewer onboard resources and can do more with less.

- The Serial class contains more parsing functions to search for data and to quickly read multiple bytes into a buffer. This may also create timing issues when using legacy code since such asynchronous operations were not available or accounted for in most sketches preceding the Arduino 1.0 release.

- Other bundled libraries like those for using the SD card reader have also been upgraded to make it easier to write sophisticated sketches without having to worry so much about the underlying code such sketches rely upon.

- Cosmetic changes have been made to the IDE. New icons, color schemes, and indicators like compilation progress bars have been added to modernize the IDE and make it easier to locate and interact with the user interface elements.

- Several other key library class and function names (such as the Wired library) have changed along with their return types and implementations. Library authors will be busy in the months ahead as they port their contributions to support these lower-level modifications.

For more details about these changes, read the entry on this topic posted on the Arduino blog.[1] Fortunately, the Arduino IDE is self-contained and portable enough to install several different versions on your computer. You will be able to continue to use the previous releases of the IDE when sketch dependencies have not yet been upgraded to support the latest improvements. As the new version becomes more widely adopted over the next year or so, more of the popular user libraries will be converted and supported. As such, future editions of this book intend to provide code compatible with the new and the old IDE releases.

Android@Home

At the 2011 Google IO conference, the Android Open Accessory API and Development Kit (ADK) was officially unveiled. The intent of this initiative was to provide Android API-level access to low-cost microcontrollers, sensors, and actuators. Conference attendees were given custom versions of the Arduino Mega board populated with basic sensors that could be polled from an Android device like the Google Nexus phone.[2] Several configuration scenarios were posited at the conference using this technology combination, one of which was dubbed Android@Home. Examples that controlled wireless lighting, entertainment systems, and exercise equipment were demonstrated, and more third-party solutions are expected to be announced at Google IO 2012.

The ADK is really what drives Google@Home, and at its heart it is a hardware specification that attempts to standardize communication across devices. The Android OS can then react to these messages accordingly. The expectation is that as hardware becomes more commoditized, the Android OS will be embedded into more devices beyond just phones. Google hopes that this will revolutionize the home automation market by having enough electronic appliance manufacturers adopt the specification and allow these devices to talk to one another.

Unfortunately, having seen this scenario play out with other home automation standardization attempts, I don't think there has been enough momentum behind the Google@Home initiative outside of Google that shows much interest…yet. Many are taking a wait-and-see approach before investing much attention. But even if Android@Home doesn't have white-hot adoption, its impact on the home automation space will no doubt spur Google's competitors, namely Apple and Microsoft, to take a closer look at this market opportunity. The most likely initial point of entry for these companies will be the television.

---

1. http://arduino.cc/blog/2011/10/04/arduino-1-0/
2. http://www.adafruit.com/products/191

The Apple Home Button

With the introduction of Apple's Siri in the iPhone 4S, Apple has constructed a meta-interface on top of information searching, one that does not rely entirely on a web browser to view query results. For search providers like Google and Microsoft, this is a game changer, since those company's revenue models are derived by interleaving relevant advertising with search results. In certain scenarios, Siri's vocal output filters these text-based results to form a conversation with the user rather than a database dump, obviating the need for a slurry of ads to be displayed. While it's technically possible that Apple may someday incorporate advertising in Siri's conversation, the near-term Siri experience is expected to be ad free. If you had the choice between typing in a query and receiving a blob of links and ads in return versus asking your TV for information and having it respond with a clear, direct answer, which technology would you use?

Apple, like Google and Microsoft, also designed a computer that connects to a television and allows streaming music and video content to be played back on the TV. Hopeful rumors abound that Apple will release a next generation version of their Apple TV device that could incorporate Siri technology for voice remote control. It isn't hard to imagine asking your TV to display the local weather forecast, play album tracks by your favorite artists, perform speech-to-text dictation email responses and, yes, even reach out to other devices in the home (predominantly iPhones and iPads) that synchronize via iCloud and participate in the conversation. Google and Microsoft won't be sitting still either, and it's possible that their voice recognition and huge data sets of aggregated information will beat Apple to the spoken command automation party.

It should also come as no surprise if engineers at Apple have been looking for ways to more tightly couple their platform into the home. With Google's Android@Home intentions and Microsoft's Kinect experiments, Apple's home consumer cards have yet to be shown. But when they are, Apple's approach will undoubtedly receive significant attention and developer support.

11.2 The Long View

While all of this automation designed around making our home lives easier is truly awesome, the one key dependency to making it all work is electricity. But you can imagine the demands placed on our planet's resources if everyone had the luxury of fully automated homes. Hopefully the next generation of entrepreneurs will do for energy collection and distribution what my generation did for computers and global communications. Smart grids, sustainable energy sources, and respect for the environment will be just as important as the inexpensive sensors, standard protocols, and ubiquitous secure wireless communication that automation products of the future will need to support.

Assuming the energy problem is accounted for, the likelihood of low-power sensors and hardware messaging systems will mushroom. How many computers, monitors, clocks, radios, phones, tablets, and entertainment consoles do you have plugged into your home's power outlets? Forty years ago, besides lighting and refrigeration, there might have been one or two TVs, an LP turntable, a few radios, and maybe an electric clock. Forty years from now, it's possible that there will be half a dozen electronic devices networked in every room and in constant chatter with their peers. Centralized services will monitor messages for events and reacting accordingly. So what will this look like?

The Home Is the Computer

Imagine taking the projects in this book and expanding them in various ways for every room in your home. Automation is everywhere and the air is busy with messages being sent to your server for processing. Perhaps this server is a virtual private server in the cloud, or maybe the message bus is being managed by a third-party provider. Your home will be able to immediately inform you of any alerts and will also be able to sense your presence and react accordingly. Image and voice recognition systems will know who you are and orient the home's services to your preferences. You will live in a sensor-filled environment and it will be just as natural and effortless as tweeting from your phone is today. The data collected will be analyzed and refined to fit your lifestyle. Your home will be capable of predicting your lifestyle activities based on external factors like the season and time of day, local weather, package deliveries, type of visitors, duration of presence, preferred mode and style of digital entertainment, and the frequency and filtering of alerts.

The Embedded Mattress

Electronic components are getting less expensive by the day. Considering how much computing power there is in a thirty-dollar Arduino board compared to the cost of the same level of computing ten years ago, it's not too difficult to imagine how even more computing capacity will be available for even less expense in the future. Combine these microcontrollers with inexpensive embedded sensors, and the home will be abuzz with information interaction. When you leave for the day, your home will power down to sleep mode, ensuring that gas and electricity consumption are kept to an operating minimum.

When you nod off to sleep, pressure sensors in your bed will know if you had a restful or restless night and accommodate the alarm in the morning. Each door could be wired so your house will know your traversal patterns and preemptively turn on lights and appliances accordingly.

For example, the house will know you wake up for work every morning at six o'clock, take a shower, and head to the kitchen for a cup of coffee thirty minutes later. After triggering your alarm clock, the shower will turn on and the water will be warm just as you enter. While you're getting dressed, coffee will be freshly brewing and ready by the time you reach the kitchen. The house will also know that you sleep in until eight o'clock on Saturdays and don't follow the same routine, so it will toggle to manual mode for the daily waking ritual. Not surprisingly, this scenario can be programmed and implemented today with the tools and technologies we used in the book's projects. But when the electronics get cheap enough, the cloud gets robust enough, and the interfaces are standardized enough, a greater number of people will come to expect this type of scenario.

11.3 The Home of the Future

Like any passionate technologist, I enjoy imagining futuristic visions of plausible technology scenarios. Yet the pragmatic developer in me knows such visions don't happen overnight. They require incremental steps in a number of related areas. But at some point, all those incremental services, discoveries, and technologies converge and create inflection points that forever alter the course of history.

I have been fortunate to participate in three major technology revolutions in my lifetime. The first was the introduction and rapid evolution of the personal computer in the 1980s. The second was the supernova expansion of the Internet in the 1990s, and the third was the mobile device revolution in the first decade of the twenty-first century.

Technologies are converging. Cloud computing, palm-sized Internet-connected supercomputers, inexpensive network-aware embedded sensors, autonomous controls, cheap storage, and faster compute cycles will lead us to another amazing era of information processing. With all these technology forces and developments meshing together, here is my prediction of a typical domestic day for a technically savvy homeowner (Figure 51, *A Smarter Home, circa 2025*, on page 199).

After coming home from a long day at work, Mel's phone activates the keyless doorway lock that automatically logs the event and video capture to her secure

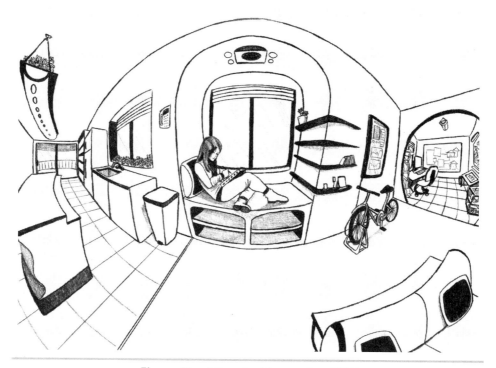

Figure 51—A Smarter Home, circa 2025

cloud bank. Based on GPS coordinates, Mel's phone had already called ahead when she was twenty minutes away to tell the HVAC system to turn on the air conditioning. By the time of her arrival, the home was as cool as when she left in the morning.

A parcel is waiting for her on the steps; it's a box of paper towels that was automatically reordered when the towel dispenser detected it was running low. With the delivery confirmation message, the dispenser's counter was automatically reset and won't need to reorder again for a while.

She's carrying a bag of groceries that her refrigerator suggested that she bring home. The sensors in the fridge detected that the tomatoes had only another day before they would start to turn, so Mel decided to pick up some additional ingredients for making spaghetti sauce.

As she prepares the meal by filling up a pot of water to boil for the spaghetti, the faucet sensors ensure that the purity of the water is contaminant-free. If an anomaly is detected, a message is sent to the city's water reclamation department automatically reporting the issue.

After dinner, Mel decides to exercise with a ride on her stationary biking simulator. She usually meets a friend on the prairie road course around this time of day, so she dons her motion-tracking 3D headset, queues up her favorite playlist and starts peddling. The headset has a heart rate, blood pressure, and perspiration monitor built into the strap, and these values are translucently overlaid on top of the pastoral scene of rolling hills of swaying wheat. After a few minutes, her friend's avatar rides close by and pings her, asking if Mel is available for conversation. Mel acknowledges and the two keep each other company as they log half an hour on their bikes. At the end of the ride, each are credited with 200 energy points as a result of their peddling power feeding electricity back into the grid.

With twilight approaching, photosensors lining the window panes prepare the home's lighting for the evening by drawing the curtains and activating motion detectors in the rooms. Gone are the days of flipping on and off light switches, unintentionally leaving lights on throughout the home even when no one is in the room. There is an override option when guests are visiting, but most of the time the motion detectors do their job well by turning on and off the lights based on presence. This effective lighting strategy has contributed to even more monthly energy credits as a result.

As she begins to settle in for the evening, Mel asks her television to list new videos that her friends have suggested. Voice control has become the norm with content consumption devices and has steadily improved with filtering algorithms and speaker identification. While the videos play back sequentially, overlays of her online status, message queues, weather forecast, and upcoming schedules can be called upon just by asking the television for that information. The weather dictates what her outfit will be the next day. A warm front is moving in, so Mel's closet rack automatically queues up via RFID sensors embedded in the clothing hangers a section of appropriate outfits to choose from. It's going to be a bright, sunny day.

I hope you enjoyed that projection of the future. For those who prefer to invent the future instead of waiting around for it to arrive, the technologies to build such a scenario exist today. With the right mix of cost-effective technology, easy implementation, and effective sales, marketing, and timing, someone is going to bring elements of this future scenario to life and forever change the way people interact with their homes. That person could be you!

CHAPTER **12**

More Project Ideas

Now that you have a solid footing on which to climb, you can take your home automation design and construction experience to new heights by building your own projects. This brief concluding chapter offers a quick survey of other ideas to consider using equipment you have already worked with.

If you assembled all the projects in this book, you already have most of the hardware required to build the ideas presented in this chapter. You can also take the code from these projects and have these new project suggestions up and running with just a few tweaks. Let's take a look at some more ways you can program your home.

12.1 Clutter Detector

Do you have a spouse, kids, or a partner who just can't seem to keep an area free of clutter and debris no matter how often you think it gets cleaned up? Somehow that empty spot attracts newspapers, junk mail, empty boxes, crumpled clothing, packing material, and whatever else happens to be pseudo-magnetized to that spot? If so, enlist the help of your newfound electronics experience by constructing a detector using an infrared distance sensor.[1]

Point the sensor at the empty space and measure the "clean area" reading. As clutter piles up, the distance detected by the sensor will be reduced. Have this trigger an email to your clutter-collecting cohabitants that they need to remove their clutter collection from the detection area. Be as aggressive as you want to on email notification frequency. And when the clutter has been removed, you can even send an email message from your home thanking the offender(s) for cleaning up the mess.

---

1. http://www.adafruit.com/products/164

12.2 Electricity Usage Monitor

Using the same concept behind Adafruit's Tweet-A-Watt,[2] hook into a Kill-A-Watt electricity flow detector to measure the energy usage of an electrical appliance such as a refrigerator or television.

Some electrical utility companies offer their customers energy tracking by month. Access this on the Web and calculate the percentage of total electricity that monitored appliance consumes on a monthly basis. Then, based on your electricity bill for that month, calculate the monthly, daily, and hourly cost associated with operating that appliance. You may be surprised just how much money you're spending on watching a couple hours of TV or how much that freezer that you bought on sale really ends up costing over its lifetime.

12.3 Electric Scarecrow

Have critter problems been plaguing your vegetable garden? Shoo those problems away using a smarter approach. Go beyond the picturesque but functionally pointless static scarecrow by bringing it to life with the help of a motion detector and a couple of heavy-duty servos. When that pesky rabbit comes by for its evening grazing on your plants, the scarecrow's sensor will bring it to life, moving its arms and legs in a convincing way to frighten the rabbit away. Have your scarecrow email a photo of its animal detection activities with the help of an Android camera phone seated inside the scarecrow's head.

12.4 Entertainment System Remote

Extend the Rails server from the *Web-Enabled Light Switch* project to transmit IR commands through a serial port connected to an Arduino that is attached to an IR LED.[3] Build the Arduino-assisted IR transmitter using Maik Schmidt's instructions in *Arduino: A Quick Start Guide* [Sch11]. Place IR LEDs in front of all your IR-controlled entertainment center devices, such as televisions and audio receivers. Create a friendly user interface for your native iOS or Android client, or combine the IR user interface with other project client interfaces, such as the one we made for the *Android Door Lock*.

---

2. http://www.adafruit.com/products/143
3. http://www.adafruit.com/products/387

12.5 Home Sleep Timer

Have any family members who fall asleep while watching TV? Forgot to power down that power-hungry quad-core desktop computer in the study? Did someone forget to turn off the lights in the basement? If so, write a script that turns off lights, appliances, and computers at a time when everyone should be sleeping. If the computer supports Wake-On-LAN (WOL), send it a shutdown packet from your script. Turn off lights via X10 heyu commands. Turn off TVs and stereos via the entertainment system remote. Turn off anything else plugged into a PowerSwitch Tail. Pocket some money, reduce your carbon footprint, and save the planet while you sleep.

12.6 Humidity Sensor-Driven Sprinkler System

Hook up a DHT22 temperature and humidity sensor to an Arduino attached to a stepper motor that drives a water spigot attached to a garden hose connected to a lawn sprinkler.[4] When the temperature is high and the humidity is low for a prolonged duration, turn on the stepper motor crank valve on the water spigot. Let the water run for ten minutes and then shut off the water valve. Calculate the volume of water used based on the duration it was running. Do this by first calibrating the number of seconds it takes to fill up a liter (or gallon, for the metric system-challenged) at the valve setting established by the stepper motor crank.

For example, if it takes thirty seconds to fill up a liter container, running the sprinkler for ten minutes will consume 20 liters (2 liters per minute times 10 minutes) of water each time you run the sprinkler. Log this amount with the help of an XBee/PC setup (from Chapter 5, *Tweeting Bird Feeder*, on page 57) over the duration of the month, and determine from your water bill the percentage of water used on your lawn. Once this metric is calibrated, you can calculate lawn sprinkling costs in real time and literally watch your money flow out of the spigot.

12.7 Networked Smoke Detectors

Smoke detectors save lives and can help minimize property damage, but what happens when the alarm goes off when nobody is home? If you know what you're doing, you can hook directly into the onboard electronics of the smoke detector to measure the voltage change when the alarm goes off, but doing

---

4. https://www.adafruit.com/products/385

so will probably void your detector's warranty. It could also put the lives of those who depend on its life-saving functionality at risk if the device is improperly modified. Instead of soldering directly onto the smoke detector's electronics, obtain an Electret microphone breakout board and use your Arduino and XBee skills to hook up, calibrate, and monitor the microphone input for the audio levels attained when the smoke detector's alarm is sounding.[5] When an alarm is detected, have your XBee message-receiving PC relay you the message via an urgent email. You could even modify the Android server we used in Chapter 9, *Android Door Lock*, on page 141, to take a photo of the area being monitored by the smoke detector and attach it to the outbound message.

You could even link this alert to perform further actions, such as auto-dialing neighbors with a recorded message asking them to investigate the fire on your behalf and call you (just in case you never received the email). And if you're really confident in your system's sensing integrity, you could go so far as to auto-call the fire department if you don't deactivate the alarm within a predetermined duration of time (though keep in mind that improper and/or nonemergency alerts could end up costing you, since many jurisdictions have penalties for false calls). Regardless of what enhancements you add, the fact remains that your smoke detector can extend its alert distance worldwide thanks to the Internet-enabled communication pathway you can devise for it.

12.8 Proximity Garage Door Opener

As you approach your garage with your GPS-enabled smartphone, the phone triggers a request to open the garage door. This relays to Arduino-XBee hardware attached to your garage door's RF-transmitting garage door opener, which in turn transmits the request to the automatic garage door receiver and opens the door.

Sans the GPS feature, opening a garage door from a smartphone like the Android or iOS device is a very popular DIY project, and a number of videos doing this have been posted on YouTube. Since your garage is a fixed location, the GPS values for latitude/longitude/elevation will remain static. By extending the Rails server we wrote in Chapter 7, *Web-Enabled Light Switch*, on page 105, writing the smartphone application that extends the toggle functionality of the garage door opening and closing based on your location shouldn't be too difficult.

---

5. http://www.sparkfun.com/products/9964

12.9 Smart HVAC Controller

Manage your air conditioning and heating needs with smarter temperature control in your home. You can dial your thermostat up or down based on a specific time frame or operate it from a remote location. Ben Heckendorn, host of *The Ben Heck Show*, posted an episode on home automation featuring this project.[6] I like his approach because he didn't mess with the hardwired internals of the thermostat. It also uses parts that we already had from Chapter 8, *Curtain Automation*, on page 125, making it an easy project to assemble and implement.

12.10 Smart Mailbox

This is another popular DIY home automation project that has numerous write-ups and video posts around the Web. Simply reuse the hardware we constructed in Chapter 5, *Tweeting Bird Feeder*, on page 57, and tweet or email when the light coming from the open mailbox lid hits the photocell. You could also have the speech playback-enabled Android device from Chapter 10, *Giving Your Home a Voice*, on page 171, audibly announce the delivery, as in "You've Got Mail!"

12.11 Smart Lighting

Go beyond the project presented in Chapter 7, *Web-Enabled Light Switch*, on page 105, to incorporate a managed lighting system throughout the home. Incorporate motion detectors to activate and deactivate lights in basements, bathrooms, and bedrooms. Record when lights turn on and off, and correlate the monthly operational costs associated with lighting your home based on your total electricity bill.

12.12 Solar and Wind Power Monitors

For those fortunate enough to have portions of their electrical power consumption supplied by residential solar and wind energy collectors, you can use the Arduino/XBee/PC combination to measure both the energy generated by these devices and the status of the battery's charge being stored in the batteries that capture and store the output of these reusable energy devices.

---

6. http://revision3.com/tbhs/homeauto

Send email alerts when the battery's charge is below a certain threshold. Capture the stats over time and map them to understand the month-to-month fluctuations that can be used to predict energy output for years to come.

If you happen to build these or any other home automation projects that you're proud of, keep the projects alive by sharing them with other readers—post your ideas, discoveries, and outcomes to the Programming Your Home book forum. See you online!

Part IV

Appendices

APPENDIX 1

Installing Arduino Libraries

One of the biggest advantages of the Arduino product line is that it is built on an open hardware platform. This means that anyone is able to contribute to the hardware and software libraries. These libraries can be easily incorporated into Arduino sketches to extend the Arduino's functionality and, in many cases, make it easier to write the sketches yourself.

Several projects in this book benefitted from such community contributions. Unfortunately, installing new Arduino libraries isn't as automatic as running a setup script. Library files, which are often distributed in a compressed .zip format, need to be uncompressed and placed into the Arduino's libraries folder. The location of this folder varies depending on which operating system the Arduino IDE is running on.

A1.1 Apple OSX

1. Locate the Arduino icon, typically found in the main /Applications folder.
2. Hold down the Control key on the keyboard and click the Arduino icon, usually located in the /Applications folder. This will pop up a context-sensitive menu.
3. Select the Show Package Contents option from the pop-up menu. This will open a folder containing the Arduino application resources.
4. Navigate to the Contents/Resources/Java/libraries folder.
5. Copy the new library files into this libraries folder.

If you prefer, you can also place library files in your home directory's Documents/Arduino/libraries folder.

A1.2 Linux

1. Locate where you uncompressed the Arduino application files.
2. Navigate to the libraries folder.
3. Copy the new library files into this libraries folder.

A1.3 Windows

1. Locate where you unzipped the Arduino application files.
2. Navigate to the libraries folder.
3. Copy the new library files into this libraries folder.

After you have copied the library files into their appropriate location, restart the Arduino IDE so that the library can be referenced in your sketches.

For example, to install the CapSense library from the *Tweeting Bird Feeder* project on a computer running Apple OS X, unzip the CapSense.zip file. Then place the unzipped CapSense folder into the /Applications/Arduino/Contents/Resources/Java/libraries folder. Restart the Arduino IDE. Create a new Arduino sketch. Type the following sketch into the Arduino IDE window:

```
#include <CapSense.h>;
void setup() {}
void loop() {}
```

Click the Verify button on the Arduino IDE toolbar. If the CapSense library was copied to the correct location, this three-line sketch should compile without errors.

APPENDIX 2

Bibliography

[Bur10] Ed Burnette. *Hello, Android: Introducing Google's Mobile Development Platform, Third Edition*. The Pragmatic Bookshelf, Raleigh, NC and Dallas, TX, 2010.

[CADH09] David Chelimsky, Dave Astels, Zach Dennis, Aslak Hellesøy, Bryan Helmkamp, and Dan North. *The RSpec Book*. The Pragmatic Bookshelf, Raleigh, NC and Dallas, TX, 2009.

[Fal10] Robert Faludi. *Building Wireless Sensor Networks*. O'Reilly & Associates, Inc, Sebastopol, CA, 2010.

[LA03] Mark Lutz and David Ascher. *Learning Python*. O'Reilly & Associates, Inc, Sebastopol, CA, 2003.

[RC11] Ben Rady and Rod Coffin. *Continuous Testing: with Ruby, Rails, and JavaScript*. The Pragmatic Bookshelf, Raleigh, NC and Dallas, TX, 2011.

[Sch11] Maik Schmidt. *Arduino: A Quick Start Guide*. The Pragmatic Bookshelf, Raleigh, NC and Dallas, TX, 2011.

[TFH09] David Thomas, Chad Fowler, and Andrew Hunt. *Programming Ruby: The Pragmatic Programmer's Guide*. The Pragmatic Bookshelf, Raleigh, NC and Dallas, TX, Third Edition, 2009.

Index

A
Adafruit Industries, 12
    Arduino Ethernet, 26
    Arduino Mega board, 195
    force sensitive resistors, 86
    infrared distance sensors, 201
    motor shield, 125, 128, 130
    music and sound add-on pack, 44
    servo motors, 44
    temperature and humidity sensors, 203
    Tweet-A-Watt, 202
    XBee adapter kits, 17, 59, 68
Android Door Lock
    Android application permissions, 162, 166
    Android client, 163–167
    Android web server, 152–158, 162
    description, 144–146
    email notification, 160–162
    hardware assembly, 146–147
    image capture, 158–160
    installation, 168
    IOIO board programming, 147–149
    parts, 144
    PowerSwitch Tail programming, 150–152
    security concerns, 164, 169
    setting static IP, 154
    testing, 167
    wiring, 147
Android OS
    Android SDK, 14–15, 107, 116
    Android test framework, 9
    application permissions, 120, 162, 166
    application signature, 122
    creating AVD, 116
    and home automation, 4, 14
    intents, 154
    vs. iOS, xv
Android smartphones, sources and costs, 13, 143
Android Virtual Devices (AVDs), 116
Android@Home
    Google Open Accessory Protocol, 142
    Open Accessory Development Kit (ADK), 14, 107, 195
Apple iOS, xv
Apple OS X, *see* Mac OS X
AppleScript, 180–188
Arduino accessories
    Ethernet shield, 26, 35–38
    motor shield, 127–128, 130–131
    MP3 shield, 46
    wave shield, 44, 46, 48–53
Arduino boards
    Arduino 1.0, 194
    Arduino Ethernet, 26
    Arduino Mega board, 195
    configuration and checkout, 27
    sources and costs, 12
    wire diagraming, 7
Arduino programming
    Arduino IDE, 15, 27, 32, 37
    Arduino: A Quick Start Guide, 27
    capacitive sensors, 62–64
    compiling and uploading, 32
    conditional blocks, 31
    defined constants, 28
    email notifications, 36–39
    end of lines, 31
    Ethernet reference library, 37
    Ethernet shield, 36–38
    flex sensors, 28–32
    force sensitive resistors, 90–92
    installing libraries, 209–210
    LED pin configuration, 29
    motor shield, 130–131
    photocells, 65–67
    PIR motion sensors, 52
    Serial Monitor Window, 32
    serial port setup, 30
    servo motors, 50–53
    stepper motors, 130, 132–136
    testing code, 9
    virtual emulators, 16
    wave shield, 50–53
    XBee radios, 70–73
Audacity, 49

B

bird feeder, *see* Tweeting Bird Feeder
Bluetooth wireless audio, 175

C

capacitive sensors, 61–64
    more project ideas, 82
CEBus standard, 4
clutter detector, 201
Curtain Automation
    Arduino programming, 132–136
    description, 128–129
    hardware assembly, 129–131
    installation, 137–139
    motor shield programming, 130–131
    parts, 127
    pulley wheels, 127
    stepper motor calibration, 138
    testing, 136

D

DIY resources
    Google Groups, xvii
    Instructables, xvii
    Ladyada, 46, 49, 65, 68, 128
    Makezine, xvii

E

Eclipse IDE, 116
Electric Guard Dog
    Arduino programming, 50–53
    description, 45
    hardware assembly, 46–48
    installation, 54
    parts, 45
    sound clips for, 47
    testing, 53–54
electric scarecrow, 202
Electric Sheep board, 145
electricity usage monitor, 202
Elektor Electronic Toolbox, 8
email notifications
    Arduino programming, 36–39
    authentication, 36
    PHP web mailer, 33–35
    using Android phone, 160–162
    using Gmail, 93, 96, 100
Emulare, 16
entertainment system remote, 202
Ethernet shield
    Arduino Ethernet, 26
    Arduino IDE reference library, 37, 194
    Arduino programming, 36–38
    assembly and wiring, 35
    DNS and DHCP functions, 38, 194
Exosite, xvi

F

FedEx tracking APIs, 95–96
flex sensors
    Arduino programming, 28–32
    more project ideas, 41–42
    wiring, 26–28
force sensitive resistors, 88
    Arduino programming, 90–92
    more project ideas, 103
    sources, 86
    wiring, 89
Freemind, 7
Fritzing, 7

G

gcc compiler, 110
Git, 31
Giving Your Home a Voice
    AppleScript programming, 180–188
    description, 171
    enabling Mac speech recognition, 175–177
    speaker setup, 173–175
    testing, 188–189
    wireless microphone calibration, 177–180
Gmail, 93, 96, 100
Google@Home, 195

H

hardware components
    costs, 12–13
    sources, 46, 59, 82
headphone-to-USB adapter, 172
Heyu utility, 105
    driving from Ruby, 114
    installing, 110
    issuing X10 commands, 111
home automation technology
    Android OS, 4
    Android@Home, 195
    CEBus standard, 4
    coming developments, 193–196
    DIY solutions, 5
    Insteon standard, 4
    programming languages, 17
    visions of the future, 197–200
    X10 system, 4
home security, *see* Electric Guard Dog
home sleep timer, 203
humidity sensor-driven sprinkler system, 203

I

iCircuit, 8
iThoughts HD, 8
IEEE 802.15.4 wireless specification, 16
infrared distance sensors, 201
infrared transmitter, 202
Inkscape, 8
Insteon standard, 4
Instructables, xvii
IOIO board, xv
    ADK support for, 14
    cost, 13
    how it works, 142
    programming, 147–149
    wiring, 146–148

L

lighting, *see* Web-Enabled Light Switch

M

Mac OS X
    AppleScript, 180–188
    Bluetooth speaker pairing, 173
    installing Arduino libraries, 209
    Mac developer tools, 110

Rails, 112
  speech recognition server, 175–177
Maker Shed, 46, 59
Makezine, xvii
microphone breakout boards, 203
microphone calibration, 177–180
microphones, wireless, 172, 177–179
miniDraw, 8
motion sensors, *see* PIR (Passive InfraRed) motion sensors
motor shield, 127
  Arduino programming, 130–131
  assembly, 128
  sources, 125
  wiring, 130
MP3 shield, 46
multimeters, 9

N
networked smoke detectors, 203

O
online resources
  Google Groups, xvii
  Instructables, xvii
  Ladyada, 46, 49, 65, 68, 128
  Makezine, xvii

P
Pachube, xvi
Package Delivery Detector
  Arduino programming, 90–92
  database, 93–95
  description, 88
  hardware assembly, 89
  installation, 102
  parts, 87
  testing, 101
  tracking number validation, 93–101
  wiring, 89
photocells
  Arduino programming, 65–67
  more project ideas, 82–83, 205
  wiring, 66, 132

PIR (Passive InfraRed) motion sensors
  Arduino programming, 52
  how they work, 49
  more project ideas, 55, 103, 140, 169, 202, 205
  wiring, 47–48
PowerSwitch Tail, 144
  programming, 150–152
  sources, 143
  wiring, 146
pressure sensors, *see* force sensitive resistors
programming languages, 17
project planning, 5–8
  components and costs, 12–17
  documentation, 9
  safety, 18
  workbench setup, 6–7
Prolific PL-2303, 109
proximity garage door opener, 204
pulley wheels, 127

R
Rails
  installation and setup, 112, 114
  server, 114, 119
RSpec, 9
Ruby on Rails, 112

S
safety considerations, 18
serial port monitoring, 69
serial-to-USB adapter, 109
servo motors
  Arduino programming, 50–53
  sources, 44
  timing adjustment, 53
  wiring, 47
Sinatra, 123
sketching, 7–8
Skype hands-free calling, 189
smart HVAC controller, 205
smart lighting, 205
smart mailbox, 205
SOAP (Simple Object Access Protocol), 95
solar power, 59, 81, 205
soldering, 6, 9

Solio, 59, 81
Sparkfun, 12
  Electret microphone breakout boards, 203
  Electric Sheep, 145
  flex sensors, 24
  force sensitive resistors, 86
  IOIO board, xv, 13–14, 142–143, 147–149
  MP3 shield, 46
  WiFly Shield, 68
SQLite, 75–76, 94
SQLite Manager, 77
stepper motors
  Arduino programming, 130, 132–136
  calibrating, 138
  how they work, 129
  more project ideas, 140, 203, 205
  wiring, 129–130
Sunforce Products, 82

T
temperature sensors
  more project ideas, 203
  wiring, 132
Test-Driven Development (TDD), 9
testing, 9
Tweet-A-Watt, 202
Tweeting Bird Feeder
  Arduino programming, 62–67
  database, 75–76
  description, 60
  hardware assembly, 61, 63, 65, 68–70
  installation, 81–82
  parts, 59
  perch sensor, 61–64
  posting to Twitter, 78–80
  seed sensor, 65
  solar power, 81
  Twitter API credentials, 77–78
  wiring, 70
  XBee radios, 68–73
Twitter interface, 77–80

U

UPS tracking APIs, 95–96

V

version control software, 31
Virtual Breadboard, 16
VirtualBox, 106
visions of the future, 197–200
voice commands, *see* Giving Your Home a Voice

W

Water Level Notifier
    Arduino programming, 28–32
    description, 25
    Ethernet shield programming, 36–38
    hardware assembly, 26–28, 35
    installation, 39–40
    parts, 25
    sending email message, 38–39
    testing, 32, 39
    web mailer, 33–35
wave shield
    Arduino programming, 50–53
    assembly and wiring, 46, 48
    sound clips for, 48
    sources, 44
    testing, 46
web access to sensors, xvi
web servers
    choices, 35
    securing notification messages, 36
Web-Enabled Light Switch
    Android application permissions, 120
    Android application signature, 122
    Android client, 115–122
    description, 108
    parts, 107
    Rails server, 114, 119
    testing, 119–122
    web client, 112–115
    X10 hookup, 109–111
WEBrick server, 114
WiFly Shield, 68
Windows
    Arduino emulators, 16
    installing Arduino libraries, 210
    VirtualBox, 106
workbench setup, 6–7

X

X10 system, 4
    connecting to computer, 109–111
    control modules, 107, 109
    Heyu utility, 105
    how it works, 108
    more project ideas, 123, 203
    potential problems, 112, 122
XBee radios
    Arduino programming, 70–73
    cost, 12
    hardware assembly, 68
    how they work, 16–17
    more project ideas, 82–83, 140, 203–204
    pairing, 68
    serial port monitoring, 69
    sources, 59

Y

Yaler, xvi

Embedded C code and Android

Whether you're working in C on an embedded platform or writing for Android phones and tablets, we've got what you need.

Still chasing bugs and watching your code deteriorate? Think TDD is only for desktop or web apps? It's not: TDD is for you, the embedded C programmer. TDD helps you prevent defects and build software with a long useful life. This is the first book to teach the hows and whys of TDD for C programmers.

James W. Grenning
(384 pages) ISBN: 9781934356623. $34.95
http://pragprog.com/titles/jgade

Google's Android is shaking up the mobile market in a big way. With Android, you can write programs that run on any compatible cell phone or tablet in the world. It's a mobile platform you can't afford not to learn, and this book gets you started. *Hello, Android* has been updated to Android 2.3.3, with revised code throughout to reflect this updated version. That means that the book is now up-to-date for tablets such as the Kindle Fire. All examples were tested for forwards and backwards compatibility on a variety of devices and versions of Android from 1.5 to 4.0. (Note: the Kindle Fire does not support home screen widgets or wallpaper, so those samples couldn't be tested on the Fire.)

Ed Burnette
(280 pages) ISBN: 9781934356562. $34.95
http://pragprog.com/titles/eband3

Arduino and The Command Line

Getting into the Arduino? Start here. And see how to make the most of command-line applications written in Ruby.

Arduino is an open-source platform that makes DIY electronics projects easier than ever. Readers with no electronics experience can create their first gadgets within a few minutes. This book is up-to-date for the new Arduino Uno board, with step-by-step instructions for building a universal remote, a motion-sensing game controller, and many other fun, useful projects.

Maik Schmidt
(296 pages) ISBN: 9781934356661. $35
http://pragprog.com/titles/msard

Speak directly to your system. With its simple commands, flags, and parameters, a well-formed command-line application is the quickest way to automate a backup, a build, or a deployment and simplify your life.

David Bryant Copeland
(200 pages) ISBN: 9781934356913. $33
http://pragprog.com/titles/dccar

Your Career, Your Blog

It's your career, learn how to make the most of it, whether you're just starting out in the field or ready to launch your latest, best blog ever.

It's your first day on the new job. You've got the programming chops, you're up on the latest tech, you're sitting at your workstation... now what? *New Programmer's Survival Manual* gives your career the jolt it needs to get going: essential industry skills to help you apply your raw programming talent and make a name for yourself. It's a no-holds-barred look at what *really* goes on in the office—and how to not only survive, but thrive in your first job and beyond.

Josh Carter
(256 pages) ISBN: 9781934356814. $29
http://pragprog.com/titles/jcdeg

Technical Blogging is the first book to specifically teach programmers, technical people, and technically-oriented entrepreneurs how to become successful bloggers. There is no magic to successful blogging; with this book you'll learn the techniques to attract and keep a large audience of loyal, regular readers and leverage this popularity to achieve your goals.

Antonio Cangiano
(304 pages) ISBN: 9781934356883. $33
http://pragprog.com/titles/actb

Pragmatic Guide Series

Get started quickly, with a minimum of fuss and hand-holding. The Pragmatic Guide Series features convenient, task-oriented two-page spreads. You'll find what you need fast, and get on with your work

Need to learn how to wrap your head around Git, but don't need a lot of hand holding? Grab this book if you're new to Git, not to the world of programming. Git tasks displayed on two-page spreads provide all the context you need, without the extra fluff.

NEW: Part of the new *Pragmatic Guide* series

Travis Swicegood
(160 pages) ISBN: 9781934356722. $25
http://pragprog.com/titles/pg_git

JavaScript is everywhere. It's a key component of today's Web—a powerful, dynamic language with a rich ecosystem of professional-grade development tools, infrastructures, frameworks, and toolkits. This book will get you up to speed quickly and painlessly with the 35 key JavaScript tasks you need to know.

NEW: Part of the new *Pragmatic Guide* series

Christophe Porteneuve
(160 pages) ISBN: 9781934356678. $25
http://pragprog.com/titles/pg_js

New Languages & New Databases

Want to be a better programmer? Each new programming language you learn teaches you something new about computing. And these aren't the databases you're used to using. Come see what you're missing.

You should learn a programming language every year, as recommended by *The Pragmatic Programmer*. But if one per year is good, how about *Seven Languages in Seven Weeks*? In this book you'll get a hands-on tour of Clojure, Haskell, Io, Prolog, Scala, Erlang, and Ruby. Whether or not your favorite language is on that list, you'll broaden your perspective of programming by examining these languages side-by-side. You'll learn something new from each, and best of all, you'll learn how to learn a language quickly.

Bruce A. Tate
(328 pages) ISBN: 9781934356593. $34.95
http://pragprog.com/titles/btlang

Data is getting bigger and more complex by the day, and so are your choices in handling it. From traditional RDBMS to newer NoSQL approaches, *Seven Databases in Seven Weeks* takes you on a tour of some of the hottest open source databases today. In the tradition of Bruce A. Tate's *Seven Languages in Seven Weeks*, this book goes beyond a basic tutorial to explore the essential concepts at the core of each technology.

Eric Redmond and Jim Wilson
(330 pages) ISBN: 9781934356920. $35
http://pragprog.com/titles/rwdata

The Pragmatic Bookshelf

The Pragmatic Bookshelf features books written by developers for developers. The titles continue the well-known Pragmatic Programmer style and continue to garner awards and rave reviews. As development gets more and more difficult, the Pragmatic Programmers will be there with more titles and products to help you stay on top of your game.

Visit Us Online

This Book's Home Page
http://pragprog.com/titles/mrhome
Source code from this book, errata, and other resources. Come give us feedback, too!

Register for Updates
http://pragprog.com/updates
Be notified when updates and new books become available.

Join the Community
http://pragprog.com/community
Read our weblogs, join our online discussions, participate in our mailing list, interact with our wiki, and benefit from the experience of other Pragmatic Programmers.

New and Noteworthy
http://pragprog.com/news
Check out the latest pragmatic developments, new titles and other offerings.

Save on the eBook

Save on the eBook versions of this title. Owning the paper version of this book entitles you to purchase the electronic versions at a terrific discount.

PDFs are great for carrying around on your laptop—they are hyperlinked, have color, and are fully searchable. Most titles are also available for the iPhone and iPod touch, Amazon Kindle, and other popular e-book readers.

Buy now at *http://pragprog.com/coupon*

Contact Us

Online Orders:	*http://pragprog.com/catalog*
Customer Service:	*support@pragprog.com*
International Rights:	*translations@pragprog.com*
Academic Use:	*academic@pragprog.com*
Write for Us:	*http://pragprog.com/write-for-us*
Or Call:	+1 800-699-7764